SpringerBriefs in Applied Sciences and Technology

PoliMI SpringerBriefs

More information about this subseries at http://www.springer.com/series/11159
http://www.polimi.it

Jacopo Maria De Ponti

Graded Elastic Metamaterials for Energy Harvesting

POLITECNICO
MILANO 1863

Jacopo Maria De Ponti ⓘ
DICA
Politecnico di Milano
Milan, Italy

ISSN 2191-530X   ISSN 2191-5318 (electronic)
SpringerBriefs in Applied Sciences and Technology
ISSN 2282-2577   ISSN 2282-2585 (electronic)
PoliMI SpringerBriefs
ISBN 978-3-030-69059-5   ISBN 978-3-030-69060-1 (eBook)
https://doi.org/10.1007/978-3-030-69060-1

This Springer imprint is published by the registered company Springer Nature Switzerland AG
The registered company address is: Gewerbestrasse 11, 6330 Cham, Switzerland

Preface

Interconnection between machines, devices and people is one of the key aspects of the contemporary society and its working paradigms, which are driving the so called *Fourth industrial revolution* or *Industry 4.0* (Hannover Fair, 2011). In this perspective, *Machine to Machine communication* (M2M) and *Internet of Things* (IoT) are able to provide increased automation, improved communication and self-monitoring, in different environments and industrial processes. A driving central force of innovation is represented by smart sensors and devices, which generate the data and allow further functionality from self-monitoring and self-configuration to condition monitoring of complex processes. This new generation of sensors has to be small, economically feasible and autonomous. The reduced power requirements of recent small electronic components make on-chip energy harvesting solutions a promising alternative to batteries or complex wiring. Amongst others, vibration-based energy harvesting schemes are particularly attractive due to the numerous and continuous sources of vibration present in the environment. However, due to the low amount of energy involved in common ambient vibrations, it is interesting to focus, or trap, waves from a larger region outside the device into a confined region in the near vicinity of the sensor. This can be obtained by exploiting the unprecedented properties of metamaterials and structuring materials in wave manipulation; once the wave is localised, by using electromagnetic, electrostatic or piezoelectric effects, efficient conversion from elastic to electric energy can be achieved.

The aim of this book is to offer a conceptual roadmap to guide the reader in the field of metamaterials and wave manipulation devices for energy harvesting applications. Two separate bodies of literature investigate acoustic/elastic metamaterials and vibration energy harvesting technologies. This book has the ambition of amalgamating these two fields inside a common framework, guiding the reader through increasing complexity. In addition, for the first time, graded multiresonator designs are proposed for energy harvesting, quantifying their advantages with respect to conventional solutions. Since this book was written single-handedly, probably contains many mistakes, and misses certain developments and contributions. To all distinguished colleagues and collaborators, I wish to present my apologies for any omissions in my text. The aim of this book is to reach a broad audience, from graduates to researchers. For this reason, it cannot be considered as an exhaustive book on both wave manipulation and energy harvesting. However, it can be useful for people approaching the world of metamaterials for the first time (first chapters), as well as researchers interested in graded metamaterial designs (last chapter).

As with every valuable research work, the results presented here would not have been possible without the continuous help and support of many people encountered in the last years. These people are dear friends, more than collaborators. A special thank goes to Dr. Gregory Chaplain, who has the merit of having formulated theoretical and analytical models for the reversed Umklapp conversion, rainbow trapping, and topological rainbow in SSH systems. He is for me a stimulating and interesting source of ideas, from whom I have much to learn. I thank Prof. Richard Craster: all this work would not have been possible without his contribution and continuous support. He is for me a great mentor, both on the scientific and human level, and this book wants to be a memory of the collaboration period spent together at Imperial College London. Last, but not least, I thank Dr. Andrea Colombi for the continuous suggestions and ideas, and Profs. Raffaele Ardito and Alberto Corigliano for their precious advice and trust my research work.

Milan, Italy   Jacopo Maria De Ponti
January 2021

Contents

Chapter 1
Introduction

Abstract In recent years the world is facing an extraordinary diffusion of the Internet of Things (IOT) concept which is the idea of building smart and autonomous sensors networks which can help us in sensing, understanding and controlling our environment. For this idea to be effective, new sensors should be small, barely costless and autonomous. Recent advances in low-power consumption circuitry have enabled ultrasmall power integrated circuits which can run with extremely low amount of power. For these reasons, the area of energy harvesting has captivated both academics and industrialists, to self power, or at least compensate, the power consumption of small electronic devices, using ambient waste energy. The integration of such systems with recent metamaterial technologies allows to dramatically increase the energy available for harvesting, and the operational bandwidth.

1.1 Preliminary Comments and Outlines

A very widespread form of energy is represented by mechanical vibrations, which exist with variable intensity in almost every environment. However, one of the main issues is that the energy involved is usually very low, and spread over a broadband low frequency spectrum. For these reasons, in order to fully take advantage of this form of energy, it is required a device that: (i) **focus or localise waves**: it is possible to increase the absorbed energy since it comes from a larger spatial region, or due to confinement in specific positions; (ii) **work in a broadband regime**: the energy in common ambient spectra can be completely used, and the performance is less affected by input changing; (iii) **can be easily manufactured**: mass scale production is possible with affordable costs.

As shown in Chap. 2, several works have been reported in literature to partially or totally address the aforementioned key requirements. Most of them rely on the design of structuring materials and metamaterials, i.e. engineered systems able to show efficient wave manipulation properties. These peculiar properties usually come from the concept of *band gap*, i.e. the existence of frequency ranges from which the propagation of waves is not allowed. For this reason, Chaps. 3 and 4 thoroughly analyse this concept in lumped systems and continuous media respectively. In Chap. 3, basic concepts on wave propagation in homogeneous and inhomogeneous media are introduced for both periodic and aperiodic structures, providing an interpretation of band gap through energy considerations and phase diagrams. Comparable attenuation capabilities are demonstrated for periodic and aperiodic structures through analytical lumped models, numerical and experimental results. Finally, a physical interpretation of the phenomenon of local resonance is provided by energy and phase considerations. This theory is generalised in Chap. 4 to elastic continua, with specific reference to plates and half-spaces. Wave propagation in thin elastic plates is firstly considered using the *Kirchhoff–Love* theory. This is followed by a numerical (using the Finite Element Method, FEM) study of wave propagation in thin elastic plates with resonators, with specific attention on the effect of the plate thickness and rod height on the band gap performance. The concept of grading is then introduced as a way to obtain broadband band gaps and the *rainbow effect*, i.e. the spatial signal separation depending on frequency. Similarly, the problem of wave propagation in elastic half-spaces is considered, going from the classical theories of Rayleigh, Shear (S) and Pressure (P) wave propagation, to the problem of the so called *metawedge*, i.e. an array of resonators able to provide rainbow effect or mode conversion depending on the direction of the incident Rayleigh wave with respect to the array grading. Finally, reversed mode conversion from surface Rayleigh to S and P bulk waves is demonstrated leveraging on the Umklapp phenomenon. This mechanism allows to manipulate surface waves, focusing the elastic energy in specific regions of space for a broadband input frequency spectrum. In this part, analytical, numerical (FEM) and experimental results for a device at the ultrasonic frequencies are reported and compared. All these concepts converge together with piezoelectric materials, to study and design piezo-augmented arrays of resonators. The choice of using as working paradigm graded arrays of resonators is motivated by their superior characteristics and versatility of use for wave manipulation, as widely shown in the recent scientific literature of the field. In addition, because such systems already contain a collection of resonators, the inclusion of vibrational energy harvesters is straightforward, leading to truly multifunctional metastructures combining vibration insulation with harvesting. First, a *rainbow reflection* device made of an array of rods with increasing height on an elastic beam is considered. By introducing a harvester inside this system, it is possible to harvest more energy with respect to the case of a single harvester to the same structure and at the same location. In addition, the harvesting bandwidth can be enlarged by introducing other harvesters in different positions. The same mechanism is also found when placing the same array of rods on an elastic half-space. In this part, analytical, numerical and experimental results are reported and compared. A second design shows that an increase of the total transduced energy along time can be obtained using *rainbow trapping* instead of *rainbow reflection*. This is demonstrated designing a graded symmetry broken array of resonators on an elastic beam, and numerically modeling (FEM) the trapping properties for energy harvesting. Finally, a third design shows that it is possible to locally increase the transduced energy by exploiting *topological edge modes* in a graded Su–Schrieffer–Heeger system. Broadband energy harvesting performances and strong robustness to impurity defects are demonstrated through numerical models (FEM) on an attractively compact device.

1.2 An Introduction to Inhomogeneous Media and Metamaterials

We introduce here the concept of *metamaterial*, as intended in this work in the setting of wave propagation phenomena. The word *metamaterial* etimologically means, from the greek prefix $\mu\varepsilon\tau\alpha$, a material with properties beyond what we expect to find in naturally occurring, or conventional materials. Unfortunately, the general nature of this definition could lead to an improper use, arriving paradoxically to say that every material, from a certain perspective, is a metamaterial. Firstly, it is important to notice that there is a strong dependance on the level at which the phenomenon is observed. In wave propagation phenomena (the ones considered here), it is reasonable to take as a reference scale of observation the *wavelength* λ, i.e. the wave spatial period. In other terms, we can adopt a material Representative Elementary Volume (REV) of the size of λ for the *homogenised* material, i.e. an homogeneous material with equivalent global properties. If λ is much larger than the smallest constitutive material element (called *unit cell*, in analogy with atoms at smaller scale) we can consider the REV as homogeneous, as typically done in continuum mechanics. If this REV shows unusual physical properties (with respect to conventional materials), we denote it as a metamaterial. Using this convention, a material is considered as a metamaterial if shows unusual properties at strong *subwavelength* scale. On the contrary, it is simply an inhomogeneous medium. In the setting of elasticity, inhomogeneous media with a periodic structure are usually called *Phononic Crystals* (PnC), coming from the term *phonon*, i.e. the quantum vibration, and *crystal* which suggest the idea of something regularly repeated in space. Specifically, the term phononic is used to say that the phenomenon involves phonons, i.e. vibrations, and that it occurs at the wavelength scale between the unit cells. The simplest example of a phononic crystal is a spring mass chain [1]. However, due to generality, the term inhomogeneous media will be adopted here instead of phononic crystal, adding the specific term of *periodic* or *aperiodic* to define the spatial arrangement of the cells. If we define the medium as inhomogeneous, we implicitly assert that inhomogeneity is a constitutive property of the material and then the wave is interacting with each unit cell, i.e. its wavelength is comparable to the unit cell size. For this reason, we adopt the term medium instead of material, since the term material belongs to something on which *average macroscopic properties* can be defined, without looking at the specific

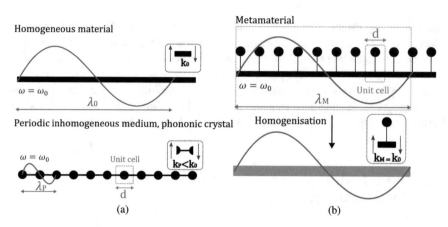

Fig. 1.1 a Homogeneous material and inhomogeneous periodic medium, i.e. phononic crystal where the wavelength (from the homogenised system) is of the order of the lattice size d. **b** Metamaterial concept: due to local resonance, an interaction with large wavelengths is possible, thus defining an equivalent homogeneous material with the same effective properties. While the stiffness of the PnC is lower with respect to the homogeneous material ($k_p < k_0 \rightarrow \lambda_p < \lambda_0$), this is not true for the metamaterial ($k_M = k_0 \rightarrow \lambda_M \approx \lambda_0$)

microstructure. Looking this in a more engineeristic rather than physical perspective, the term *structure* can be adopted, meaning an assembly of subelements, represented by the unit cells. Figure 1.1a compares a homogeneous material and inhomogeneous periodic medium (PnC). Even if they are made of the same material, the wavelength associated to the same frequency is remarkably different, mainly due to the different global stiffness. On the other hand, Fig. 1.1b shows the concept of metamaterial. The addition of resonators on the homogeneous material maintains the same global stiffness, thus slightly changing the wavelength. In this way, the system can be regarded as nearly homogeneous. Since the mechanical concept of continuum is meaningful if the behaviour is strongly subwavelength (Fig. 1.1b), metamaterials can be based in essence only on resonance effects, in accord with [2] (this mechanism will be explained in detail in this chapter).

This is coherent with the seminal work of the group of Ping Sheng at HKUST [3], that provided the first numerical and experimental evidence of a localised resonant structure for elastic waves propagating in three-dimensional arrays of thin coated spheres. Adopting this interpretation, metamaterials, contrary to phononic crystals, can be even aperiodic. However, they are usually periodically defined, due to the peculiar properties given by periodicity, as well as the reduced computational complexity and the existence of analytical closed form solutions.

The work of Liu in acoustics opened the door to the design of elastic metamaterials, but this concept was preceded by important discoveries in electromagnetism and optics. In 1967, the Russian physicist Victor Veselago published a visionary paper [4] in which electromagnetic media with simultaneously negative permittivity ε and magnetic permeability μ were shown to be characterized by a negative refractive

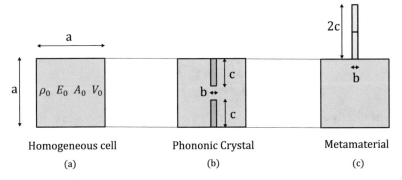

Fig. 1.2 **a** Homogeneous cubic cell with volume V_0, density ρ_0 and elastic modulus E_0. **b** Phononic crystal cell obtained removing from the omogeneous one a volume equal to $2abc$. This procedure completely changes the global stiffness of the structure, thus defining a smaller wavelength for the equivalent homogeneous cell, i.e. the scale of observation is reduced. **c** Metamaterial obtained adding a volume $2abc$ to the homogeneous cell. Even if this procedure changes the global mass of the system, this effect on the wavelength, i.e. the scale of observation, can be very small

index of refraction. He showed that a slab of a negative refractive index material can act as a flat convergent lens that images a source on one side to a point on the other. This discovery remained an academic curiosity for almost three decades, until the British physicist John Pendry [5, 6] proposed effective designs of structuring materials with negative ε and μ, and experimental demonstrations were done at the GHz frequencies by a handful of photonic groups in the United States (2000) [7]. As previously emphasised, these materials are structured at subwavelength scale (typically $\lambda/10$), hence it is possible to regard them as nearly homogeneous. The term metamaterial, coined by Pendry, describes such periodic structures when one can *average* their properties, which are strongly dispersive and anisotropic [2]. The paradigm of potential applications is exemplified by Pendry's perfect lens which involves not only the propagating wave, but also the evanescent near-field component [8], ensuring a resolution beyond the Rayleigh criterion.

Before concluding this introductory section on the metamaterial concept, we show a simple example to heuristically clarify the proposed definition. To do so, we anticipate some basic equations fully demonstrated in the next chapter. Let us consider a cubic homogeneous cell, as the one shown in Fig. 1.2a.

The total volume is $V_0 = a^3$, while the mass and longitudinal (axial) stiffness are $m_0 = \rho_0 V_0 = \rho_0 a^3$ and $k_0 = E_0 A_0/a = E_0 a$ respectively. Using the relation between frequency and wavenumber (Eq. (3.7)) explained in the next chapter, we obtain: $\lambda_0 = (1/f_0)\sqrt{E_0/\rho_0}$. We consider now a Phononic crystal cell, obtained removing a volume equal to $2abc$ from the homogeneous one (Fig. 1.2b). The total mass is $m_p = \rho_0 V_P = \rho_0(a^3 - 2abc)$ while the longitudinal stiffness, approximated by considering the smallest element only, is $k_p = E_0 a(a - 2c)/b$. We now assume to define for this cell, an equivalent one with volume V_0, but mass and stiffness equal to the actual one. This defines an equivalent density and elastic modulus as

$\rho_p = m_p/V_0 = \rho_0(a^2 - 2bc)/a^2$ and $E_p = k_p/a = E_0(a - 2c)/b$. The wavelength corresponding to the same frequency f_0 is thus given by:

$$\lambda_p = \frac{1}{f_0}\sqrt{\frac{E_p}{\rho_p}} = \lambda_0\sqrt{\frac{a^2(a - 2c)}{b(a^2 - 2bc)}}. \tag{1.1}$$

It can be clearly seen that for $c \to (a/2)$ the wavelength $\lambda_p \to 0$, i.e. the scale of observation of the problem becomes infinitely small, thus defining an inhomogeneous medium. Let us now assume to add the mass removed to define the Phononic crystal, to the initial homogeneous cell (Fig. 1.2c). Using the same approach, we obtain a total mass $m_M = \rho_0 V_M = \rho_0(a^3 + 2abc)$ and stiffness $k_M = E_0a$. The equivalent volume with equal global mass and stiffness, has density $\rho_M = m_M/V_0 = \rho_0(a^2 + 2bc)/a^2$ and elastic modulus $E_M = E_0$. The wavelength corresponding to the same frequency f_0 is thus given by:

$$\lambda_p = \frac{1}{f_0}\sqrt{\frac{E_M}{\rho_M}} = \lambda_0\sqrt{\frac{a^2}{a^2 + 2bc}}. \tag{1.2}$$

It can be noticed that for $c \to (a/2)$ the wavelength $\lambda_p \to \sqrt{a^2/(a^2 + 2b)}$, i.e. the scale of observation of the problem remains finite. In addition, for the limit of $b \to 0$, $\lambda_M = \lambda_0$.

References

1. L. Brillouin, Wave propagation in periodic structures: electric filters and crystal lattices. Nature (1946)
2. R.V. Craster, S. Guenneau, *Acoustic Metamaterials: Negative Refraction, Imaging, Lensing and Cloaking* (Springer Nature, New York, 2013)
3. Z. Liu, X. Zhang, Y. Mao, Y.Y. Zhu, Z. Yang, C.T. Chan, P. Sheng, Locally resonant sonic materials. Science (2000)
4. V.G. Veselago, The electrodynamics of substances with simultaneously negative values of ϵ and μ. Sov. Phys. Uspekhi (1968)
5. J.B. Pendry, A.J. Holden, W.J. Stewart, I. Youngs, Extremely low frequency plasmons in metallic mesostructures. Phys. Rev. Lett. (1996)
6. J.B. Pendry, A.J. Holden, D.J. Robbins, W.J. Stewart, Magnetism from conductors and enhanced nonlinear phenomena. IEEE Trans. Microw. Theory Tech. (1999)
7. D.R. Smith, W.J. Padilla, D.C. Vier, S.C. Nemat-Nasser, S. Schultz, Composite medium with simultaneously negative permeability and permittivity. Phys. Rev. Lett. (2000)
8. J.B. Pendry, Negative refraction makes a perfect lens. Phys. Rev. Lett. (2000)

Chapter 2
State-of-the-Art of Engineered Materials for Energy Harvesting

Abstract This chapter contains a brief presentation of structuring materials and metamaterials proposed in literature for vibration-based energy harvesting. The following systems are considered: Parabolic Acoustic Mirror (PAM), Resonant defect inside a Phononic Crystal (PnC), Acoustic Funnel, Multifunctional device based on local resonance, Lensing, Acoustic Black Hole (ABH), and graded designs.

Metamaterials, and in general structuring materials, due to their unique properties to guide the propagation of elastic waves and focus their energy, can be adopted to enhance vibration based energy harvesting [1]. These features have been exploited for the design of innovative actuators and sensors, and elements of logic circuitry based on the propagation of elastic waves [2, 3]. By operating within or close, to a band gap (i.e. a frequency range for which the wave propagation is forbidden) waves can be guided and focused. We investigate now the main approaches proposed in literature to focus elastic energy using structuring materials or metamaterials, usually combined with smart materials for energy harvesting or sensing. For an overview of the fundamental concepts of wave physics (here assumed known), the reader is referred to the next chapter.

2.1 Parabolic Acoustic Mirror (PAM)

An Elliptical Acoustic Mirror (EAM) made of short rods on a plate (i.e. phononic crystal based) [4] represents an effective means to focus propagating waves. However, it has the drawback that the source location should be known a priori, which is almost impractical for most of real-life scenarios. A *Parabolic Acoustic Mirror (PAM)* is able to overcome this limitation, focusing plane waves from an unknown

J. M. De Ponti, *Graded Elastic Metamaterials for Energy Harvesting*,
PoliMI SpringerBriefs, https://doi.org/10.1007/978-3-030-69060-1_2

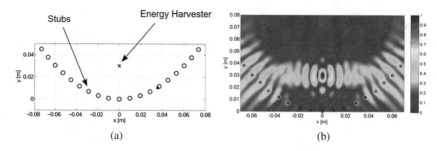

Fig. 2.1 **a** Schematic of the PAM made of short pillars on a plate with the piezoelectric energy harvester; **b** detail of the simulated RMS displacement distribution exhibiting focusing of the wave energy at the location of the energy harvester [1]

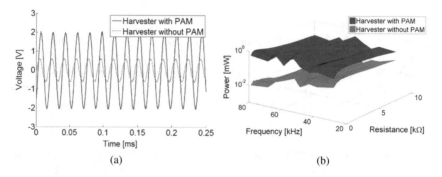

Fig. 2.2 **a** Voltage output histories with/without the PAM at 55 kHz for 1.3 kΩ exhibiting the advantage of the PAM-based energy harvester; **b** power versus load resistance and frequency surfaces for the frequency range 30–80 kHz [1]

source, as shown in Fig. 2.1 from [1], for a plate with short pillars. The performance of the PAM-based wave energy harvesting concept proposed in [1] is investigated experimentally and compared to that of a free wave energy harvester. Figure 2.2a shows the increment of the voltage output in the PAM-based harvester as compared to the free energy harvester without the PAM. Strong enhancement of the energy harvesting capabilities of the PAM-based configuration are shown in Fig. 2.2b, where a substantially enhanced power generation performance over a wide range of excitation frequencies is obtained. Specifically, the maximum power generated by the PAM-based energy harvester occurs at 55 kHz and 1.3 kΩ with 1.51 mW, whereas the free harvester showed a maximum at 55 kHz and 1.7 kΩ producing 145 μW of power.

In the same work, the authors propose energy harvesting enhancement by introducing a *defect* inside a 2D lattice made of aluminium stubs on a plate, as done years before in acoustics [5].

2.2 Resonant Defect Inside a Phononic Crystal (PnC)

This concept has been widely proposed in both elasticity and acoustics [5–10]. The main idea is to create suitable resonant defects inside a phononic crystal to confine the strain energy originating from an acoustic or elastic incident wave. In [8] the authors show that a localised mode in a defected PnC can be used for acoustic energy harvesting taking advantage from the high local strain energy density. Figure 2.3a shows a schematic of the proposed acoustic energy harvesting system made of a defected supercell with a piezoelectric PZT patch connected to a resistive electric load. The high strain energy density of the defect mode (see Fig. 2.3b) allows to enhance the harvestable energy.

By inspecting the dispersion curves of the defected supercell with the PZT patch, a flat defect mode with frequency around 2254 Hz (Fig. 2.4a) can be noticed inside the acoustic band gap. This opens a gate of transmission in the gap and provides the possibility of energy confinement.

The authors further validate the acoustic model by computing the Sound Transmission Loss (STL) curves of the system with and without the defect (see black and blue lines in Fig. 2.4b), forced with an acoustic plane wave incidence with a sound pressure of 2 Pa (100 dB). It can be noticed in Fig. 2.4b that the defect mode (see the blue curve with solid circles) at the frequency of 2257.5 Hz shows high sound transmission (STL $\approx$ 0) in the band gap, while the transmission maintains low under no defect condition (refer to the black line). By using defect modes it is possible to achieve strong energy harvesting enhancement, but a narrowband source and a direct excitation inside the defect is required. As the periodic structure is able to confine waves inside the defect, it scatter them when they come from an outside region at different frequencies, which is a limitation for realistic applications.

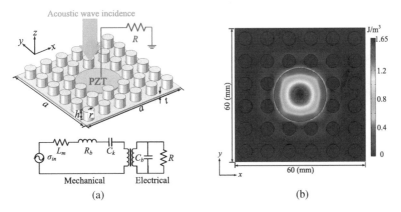

Fig. 2.3 **a** Schematic of the acoustic energy harvesting system composed of a defected supercell with a piezoelectric patch and a load circuit with the equivalent circuit representation. **b** Strain energy density distribution at the defect mode at 2257.5 Hz with 2 Pa acoustic wave [8]

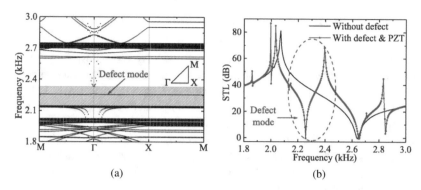

(a)   (b)

Fig. 2.4 a Dispersion curves computed by FEM for the defected supercell with piezoelectric patch and **b** transmission loss with and without the defect [8]

2.3 Acoustic Funnel

Another usually proposed approach is based on the creation of an *acoustic funnel* by means of acoustic scatterers on a plate, as defined in [1]. The operating principle relies on the capability of such a periodic arrangement to manifest band gaps, based on the periodic spacing of the aluminum stubs. Waves at these frequencies do not propagate through the stubbed portion of the plate and are bounded in the bare plate region. Such region is shaped to capture circular crested waves generated by a piezoelectric source, acting as a point source. The funnel effectiveness is measured taking as reference a free harvester condition outside the funnel, as shown in Fig. 2.5a.

The average power dissipated across the resistor (which is the AC power produced by the piezoelectric energy harvester disk) against the resistance (Fig. 2.5b) shows that the optimal electrical load is the same for the two cases. In addition, the

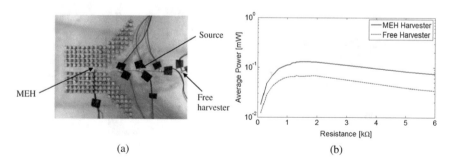

(a)   (b)

Fig. 2.5 a Acoustic funnel formed by the periodic arrangement of stubs featuring an open channel along which waves are guided. **b** Comparison of power generation performance for the acoustic funnel and the free harvester configuration [1]

peak power jumps from 67 to 130 μW, showing the energy harvesting enhancement provided by the funnel configuration.

2.4 Multifunctional Device Based on Local Resonance

Parallel to the energy harvesting enhancement due to mirroring, guiding and localization, comes the idea to design multifunctional devices. This is an immediate consequence of the band gap concept, which simultaneously allows for energy localization in specific regions of space, and attenuation in others. For this reason, multifunctional structural designs able to simultaneously combine superior mechanical wave filtering properties and energy harvesting capabilities have been widely proposed in literature. A natural application of these properties can be found for the problem of insulation of sensitive self-powered electronic microsystems from environmental and engine-generated structural vibrations on board of space systems [11]. The concept is schematically illustrated in Fig. 2.6, as explained in [12]. The device is intended to work as a multifunctional vibration absorber separating two environments A and B, where it is assumed that A is subjected to externally applied mechanical loads and B needs to be powered and insulated from potentially harmful excitations. Multifunctionality is achieved by designing the core of the device as a periodic structure, able to exploit band gaps and internal wave localization [13–16]. Due to the band gap, flat regions with zero group velocity are available immediately below and above the gap itself. This means that the vibrational kinetic energy localizes in the form of an oscillatory motion of the internal structural elements, rather than being transferred across the material as propagating waves.

The idea is to add piezoelectric materials in the substructures, to transduce the localised kinetic energy in the resonators into electrical one. Figure 2.7a shows a

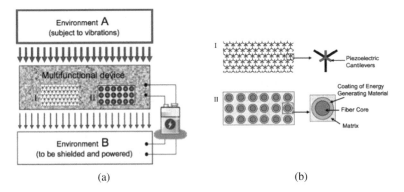

(a)   (b)

Fig. 2.6 **a** Multifunctional device for vibration isolation and energy harvesting, introduced between an environment A (subject to vibrations) and B (to be shielded and powered). **b** Internal configurations with the insertion of smart materials [12]

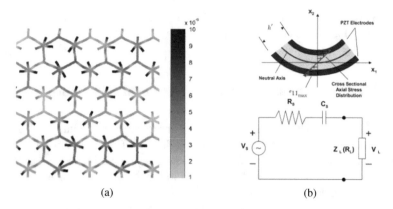

Fig. 2.7 **a** Displacement field in a honeycomb lattice of the region with extreme localized deformation at the band gap edge and **b** beam with piezoelectric patches with equivalent electric circuit representation [12]

zoomed view of the localised deformations in the honeycomb lattice proposed in [12], while Fig. 2.7b the corresponding deformation of the piezoelectric patches with a schematic of the circuit.

A similar concept is presented in [17, 18], while in [19], spring-loaded magnets are adopted in a resonator lattice system instead of piezoelectric materials. In other works [20–22], compact membrane based acoustic metamaterials are proposed, able to show simultaneously high sound transmission loss and efficient energy conversion using piezoelectricity or magnetic transduction [23]. Other approaches are based on the design of coiled configurations [24, 25] to achieve the desired phase profiles for the acoustic focusing and energy confinement. For common applications involving low frequency spectra, local resonance is the most suitable mechanism. In [26] a novel sub-wavelength scale energy scavenger able to harvest energy at frequencies lower then 1 kHz is designed using a spherical heavy core made of lead, encapsulated inside a matrix endowed with a piezoelectric wafer. In [27] a similar local resonance mechanism is adopted, proposing also a multi-cell metamaterial model for broadband behaviour. Figure 2.8a shows the unit cell adopted in [27], made of a rectangular aluminum frame containing a cylindrical matrix with an encapsulated spherical heavy core of lead and a piezoelectric wafer. For negative dynamic effective mass, the wave energy is trapped inside the soft matrix, corresponding to the dominant electric power peaks. Output electric power values up to 35 μW are obtained for an electric load of 10 kΩ. Figure 2.8b shows the displacement patterns of the unit cell for different frequencies, with the corresponding excitation of the piezoelectric wafer. Since the dynamic effective mass of the system strongly depends on the mass of the core resonators, the same authors propose a multi-frequency scavenging model, made of multiple cells with linearly varying core masses. Figure 2.9a shows multiple peaks in the normalised voltage depending on frequency, each one related to a different resonance of the internal masses (Fig. 2.9b).

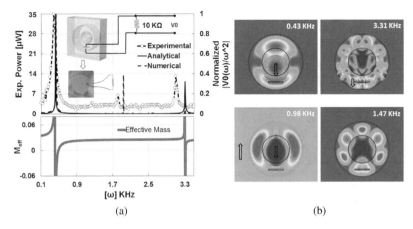

(a)   (b)

Fig. 2.8 **a** Analytical, numerical and experimental output power against 10 kΩ electric load, and dynamic effective mass. **b** Displacement patterns in the unit cell at 0.43, 3.31, 0.98 and 1.47 kHz [27]

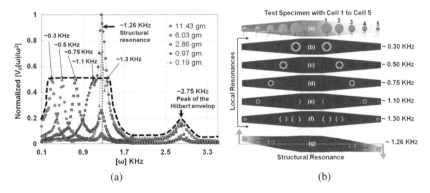

(a)   (b)

Fig. 2.9 **a** Normalised experimental power output using a multi-cell metamaterial and **b** corresponding displacement patterns at 0.30, 0.50, 0.75, 1.10, 1.30 kHz, and structural resonance at 1.26 kHz [27]

In [28, 29] it is demonstrated that useful energy can be harvested using mechanical and electromechanical locally resonant metastructures [30, 31]. Figure 2.10a shows the proposed mechanical metastructure where small cantilever beams with tip masses act as mechanical resonators attached to the primary beam structure. Similarly, an electromechanical metastructure is proposed using a piezoelectric bimorph with segmented electrodes with resonators defined by inductors (Fig. 2.10b).

For what concerns the mechanical metastructure, it is shown in Fig. 2.11 that large, relatively broadband power output occurs near the resonance frequency of the resonators, just before the locally resonant band gap, approximately defined as [29]:

$$\omega_t\sqrt{1+\gamma} < \omega < \omega_t\sqrt{1+\mu+\gamma}, \qquad (2.1)$$

(a)   (b)

Fig. 2.10 **a** Mechanical locally resonant metastructure for energy harvesting, made of cantilevers on a primary beam with piezoelectric patches attached to a resistive load. **b** Electromechanical locally resonant metastructure. The primary structure is a piezoelectric bimorph with segmented electrodes. Inductors shunted to each pair of electrodes serve as electromechanical resonators, and resistors are placed in parallel to provide energy harvesting capability [29]

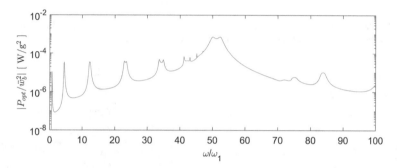

Fig. 2.11 Optimal real power output for the mechanical metastructure with resistive shunting with respect to the normalized excitation frequency ω/ω_1. Mass ratio $\mu = 1$, dimensionless coupling term $\gamma = 0.1$, short circuit resonance frequency $\omega_t = 50\omega_1$ and damping rations $\xi = 0.01$ and $\xi_r = 0.01$ [29]

where ω_t is the resonance frequency of the resonators, γ a dimensionless piezoelectric coupling term and μ the ratio between resonating and non resonating mass. The strong attenuation inside the band gap (Fig. 2.12) is not affected by the presence of the piezoelectric patches with the associated harvesting circuits, demonstrating that energy harvesting and vibration isolation can be well integrated. Additionally, the resonance frequencies before the band gap are strongly attenuated by the harvesting circuit. Thus energy harvesting in a mechanical metastructure can help attenuate the resonances before the locally resonant band gap while providing useful power for sensing and other applications.

The authors also show that the optimal power output for the electromechanical metastructure displays more broadband behavior but less power than the mechanical metastructure (Fig. 2.13), likely due to the narrow-band nature of mechanical resonator type energy harvesters.

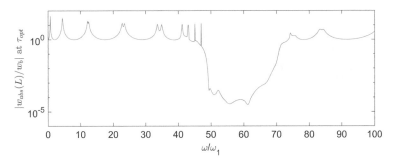

Fig. 2.12 Beam tip response for the mechanical metastructure at optimal power output with respect to the normalized excitation frequency ω/ω_1. Mass ratio $\mu = 1$, dimensionless coupling term $\gamma = 0.1$, short circuit resonance frequency $\omega_t = 50\omega_1$ and damping ratios $\xi = 0.01$ and $\xi_r = 0.01$ [29]

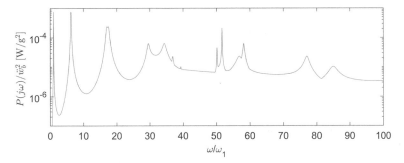

Fig. 2.13 Optimal real power output for the electromechanical metastructure with resistive shunting with respect to the normalized excitation frequency ω/ω_1. Electromechanical coupling term $\alpha = 0.44$, short circuit resonance frequency $\omega_t = 50\omega_1$ and damping ratio $\xi_r = 0.01$ [29]

2.5 Lensing

Another approach widely proposed in literature is based on the creation of lenses able to focus the elastic energy in specific regions of space. In [32] enhanced elastic energy harvesting is obtained designing a Gradient-Index Phononic Crystal Lens (GRIN-PCL) [33–38] made of an array of holes with different diameters in an aluminium plate. Figure 2.14a shows the unit cell adopted in [32], obtained introducing a blind hole in an aluminium plate. By changing the size of the hole, it is possible to modify the dispersion curves, obtaining different slopes for different filling factors. Adopting an hyperbolic secant profile of the refractive index, it is possible to focus waves in specific positions. The authors quantify the energy harvesting performance of the lens by bonding piezoelectric disks at the first focal point in the GRIN-PCL and also in a baseline setting at the same distance from the excitation source in the uniform plate region.

Figure 2.14b shows the voltage under optimal resistive load of 2.2 kΩ and the average power across a set of resistive loads for a 50 kHz sine burst excitation. Under the

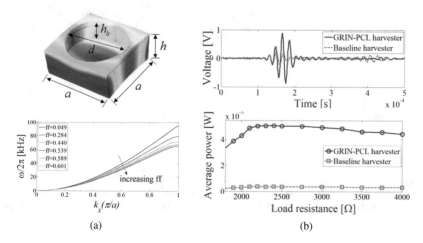

Fig. 2.14 **a** Unit cell structure of the GRIN-PCL made of a plate with blind holes, and band structure of the A_0 Lamb mode for various filling factors ($ff = \pi d^2/4a^2$). **b** Experimental voltage response histories of the GRIN-PCL harvester and the baseline harvester under optimal resistive loading (2.2 kΩ), and average power output for a set of resistive loads for a 4-cycle 50 kHz sine burst excitation [32]

same excitation applied to both harvesters, the efficiency is increased by 13.8 times by focusing the elastic waves in the GRIN-PCL as compared to the baseline case of harvesting incident plane waves using an identical piezoelectric disk without the lens. Similar results are obtained in [39] by varying continuously the plate thickness. Figure 2.15a shows the numerical mean-squared integral of the wavefield velocity, with two focuses at a distance $L_f = 3.2$ cm from each other. Similarly to [32] the energy harvesting performance is experimentally measured introducing piezoelectric disks and comparing the focal and baseline harvester. Figure 2.15b shows the experimentally measured average power output generated by the piezoelectric disks versus load resistance. The generated power obtained with the focal harvester is two orders of magnitude larger than the baseline power.

In [40] a phononic crystal Luneburg lens is designed for omnidirectional elastic wave focusing and enhanced energy harvesting. The lens is obtained by means of hexagonal unit cells with blind holes of different diameters, determined according to the Luneburg lens refractive index distribution related to the A_0 Lamb mode. While the GRIN-PCL [32] is susceptible to the orientation of the incident plane wave, the PnC Luneburg lens [40] is able to provide omnidirectional focusing. In order to quantify the energy harvesting capabilities, piezoelectric energy harvesters are bonded at the edges of the lens domain and in a baseline setting, for focal regions of 0° and 30°, as shown in Fig. 2.16a. Under the same excitation applied to both harvesters, the harvested power is increased of around 13 times by focusing the elastic waves in the PC Luneburg lens with respect to the baseline case (Fig. 2.16b).

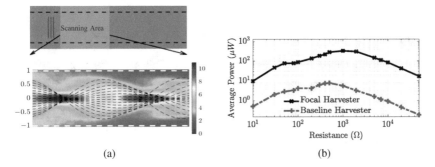

(a)   (b)

Fig. 2.15 **a** Numerical nondimensionalised mean squared integral of velocity generated by line source excitation at 40 kHz in a continuous flexural GRIN lens. **b** Average power produced for different electric loading for lens and baseline configuration [39]

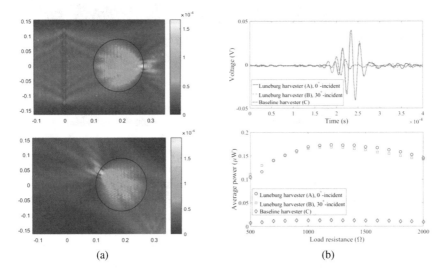

(a)   (b)

Fig. 2.16 **a** Experimental RMS wave fields demonstrating the Luneburg lens focusing for $0°$ and $30°$ angles of incidence. **b** Experimental voltage under the optimal resistive loading of 1.2 kΩ, and average power with changing load resistance for the Luneburg and baseline harvester [40]

2.6   Acoustic Black Hole (ABH)

Other approaches are based on the creation of Acoustic Black Holes (ABHs) [41, 42]. An ABH can be regarded as the acoustic counterpart of the optical one, i.e. an object from which light cannot escape. This is obtained by gradually varying the thickness of a plate or bar [43], through a power law relationship between the local thickness and the distance of the edge, with the form $h(x) = \varepsilon x^{m}$, where ε is a constant and $m \geq 2$.

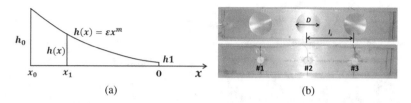

Fig. 2.17 **a** Schematic of an ABH defined through a wedge with power law profile and residual thickness h_1. **b** Prototype composed of three ABHs, with surface mounted piezoceramic transducers (PZT), in a cantilever aluminium plate [42]

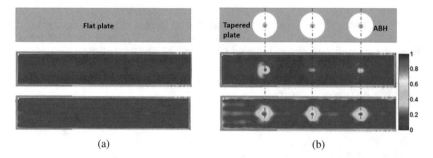

Fig. 2.18 Prototype schematic and RMS of the displacement field for the flat (**a**) and tapered (**b**) plate with ABHs at the excitation frequency of $f_{c1} = 10$ kHz and $f_{c2} = 20$ kHz [42]

The gradual stiffness reduction results in a smooth decrease of the phase and group velocities of the elastic waves, enabling a minimization of the edge reflections [44]. However, due to deviations of real manufactured wedges from the ideal power law shapes, largely due to the unavoidable truncations of the wedge edges (see Fig. 2.17a), high reflections exist [43]. This effect can be reduced by covering the sensitive parts of the wedge surfaces by thin damping layers [45–49], obtaining reflection coefficients of the order of 1–3% [47]. Alternatively to initial applications based on the minimization of edge reflections, the capability of the ABH to localise and trap waves can be also used for energy harvesting. In [41, 42] this is studied numerically and experimentally by creating a plate with ABHs and Piezoceramic Transducers (PZT) bonded on the flat surface of the tapered plate and centered at each ABH, as shown in Fig. 2.17b. The introduction of ABHs provides strong wave localization and amplification (Fig. 2.18a, b) with respect to the bare plate.

The lower frequency f_{c1} is almost entirely slowed down by the first ABH, while at the higher frequency f_{c2}, more energy can pass through the tapers (due to the higher initial velocity), therefore producing higher energy accumulation at the following ABHs (Fig. 2.18b). The effect of the ABHs for energy harvesting is studied experimentally [42] both in steady state and transient regime, obtaining respectively an increase of 400 and 250% of the harvested energy with respect to the bare plate (Fig. 2.19).

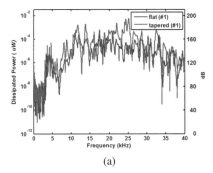
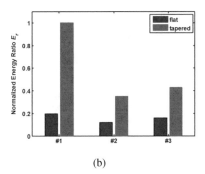

(a)   (b)

Fig. 2.19 Experimental results on energy harvesting performance under steady state response. **a** Dissipated power between the flat and tapered plates at $PZT\#1$ and **b** cumulative normalised energy ratio over the frequency range 0–40 kHz for the three transducers [42]

Specifically, this comparison is obtained computing the non-dimensional normalised cumulative energy ratio $E_r = E_0^{(n)}/E_i max_n[E_0^{(n)}/E_i]$, where $E_0^{(n)}$ is the energy dissipated by the resistor connected to the nth transducer, and E_i is the mechanical input energy.

2.7   Graded Designs: The Rainbow Effect

Another approach for the wavefield amplification relies on the *rainbow effect*, i.e. the spatial signal separation depending on frequency, firstly studied in electromagnetism [50], and then in acoustics [51–55] and elasticity [56–58]. By gradually varying the medium effective properties, it is possible to modify the waves, that are spatially compressed and amassed, with a strong amplitude enhancement [50, 51, 54, 55, 59]. In [53] an anisotropic acoustic metamaterial made of an array of stainless steel plates spaced by air gaps is proposed to have strong amplification of the pressure fields.

Similarly, sound enhancement was demonstrated using adiabatic linear [54] and exponential lattice spacing variation [55] in chirped crystals. Figure 2.20a shows the dispersion curves and spatial wave profile for the exponential chirped structure proposed in [55]. By grading the system, it is possible to change the band gap within the discrete structure. Specifically, as the filling fraction gradually increases along the chirped crystal, the width of the band gap also gradually increases. Due to this, the wave entering into the crystal is gradually slowing down, approaching during propagation the edges of local band gap inside the crystal. At a particular depth corresponding to the band-edge, where the group velocity is zero (in absence of losses), the forward propagating wave stops, turns around, and starts propagating backwards, suffering a "soft" reflection, as shown by the spatial profile in Fig. 2.20a.

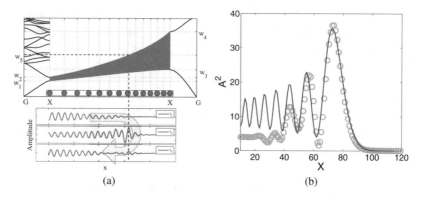

Fig. 2.20 a Local band gaps in exponential chirped structures (top) and spatial profile of a pulse evaluated at an instant t_1 before the pulse reaches the turning plane, at the instant t_2 when the pulse reaches the turning plane and at the instant t_3 after the pulse reaches the turning plane. **b** Analytical (solid line) and experimental (green dots) amplitude profile in the exponentially chirped crystal at 2700 Hz [55]

A coupled mode theory is proposed to analytically model the amplitude profile, obtaining a good prediction of the experimental results.

Analytical and experimental results confirm pressure enhancement, as shown in Fig. 2.20b for an input at 2.7 kHz. In addition to this, a time spreading of the signal in the chirped crystal is also measured. Figure 2.21 shows the time-space scenario at $f = 2.7$ kHz. Whereas the recorded signal for the case of a rigid wall is the superposition of incident and reflected waves with a determined duration, the recorded signal for the chirped structure has a longer duration due to the slowing down of the group velocity and it is modulated by the several contributions due to multiple reflections back and forth of the wave inside the chirped crystal. Moreover, the character of the sound enhancement depends on the function of the variation of dispersion, i.e. on the function of the chirp. Specifically, a stronger sound enhancement is obtained for the exponentially rather than linearly chirped crystals [55]. Parallel to the sound enhancement in the position in which the frequency components accumulate, the input time spreading (Fig. 2.21b, d) offers enormous advantages for energy harvesting since more energy can be stored along time. In the setting of elasticity, the rainbow effect has been demonstrated for Rayleigh surface waves interacting with an array of resonators with increasing height, both for seismic [57] and ultrasonic [58] applications. This system, defined as *metawedge*, is able to provide rainbow effect or mode conversion depending on the direction of the incident Rayleigh wave with respect to the grading of the array. For the increasing height metawedge, surface waves are slowed down inside the metawedge and their wavelength become as short as the spacing between the resonators. Contrary, for the decreasing height metawedge, Rayleigh waves are mode converted into shear waves. Figure 2.22 shows the experimental results in [58] for an array of resonators obtained through micro-milling of an aluminium block.

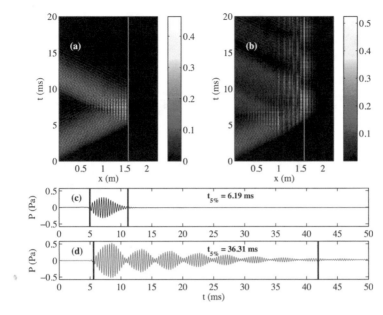

Fig. 2.21 Time spreading of the input signal in a chirped sonic crystal. **a, b** show the time signals recorded for a longitudinal cut in x direction, for an homogeneous medium (air), and the exponential chirped structure placed in the range $x = [1, 2]$ m, respectively. A rigid wall is placed in $x = 1.765$ m for the homogeneous case. Green solid lines indicate the position of the time signals shown in **c**, **d**. Recorded time signals for the homogeneous medium (**c**) and the exponential chirped structure (**d**), at $x = 1.505$ m [55]

The metawedge consists of a 1.5 mm spaced array of 40 rows each containing 18 microresonators with wedge height varying with an angle $\theta \approx 2.5°$ from 3 mm the tallest to 0.5 mm the shortest. Figure 2.22a shows the *classic* metawedge case. By looking at the top surface, spatial signal separation depending of frequency is observed, as well as amplitude amplification at the surface and in the resonators. For the inverse metawedge in Fig. 2.22b the inspection of the bottom surface reveals strong signals produced by the converted Rayleigh waves into shear waves. Both cases show a good agreement with theoretical predictions based on the spatial variation of the longitudinal resonance and Snell's law. The physics of this system is described through a Fano-like resonance [60], where the coupling between the rods and the Rayleigh waves at the longitudinal resonances creates large band gaps. Mathematically, the eigenvalues of the equations describing the motion of the substrate and the rod are perturbed by the resonance, complex roots arise and form a band gap [61, 62]. As for the acoustic case, the rainbow effect yields large amplifications, more specifically in the resonators where the wave is stopped and then reflected back. This phenomenon is explained in detail in the next chapters, since it is at the base of the energy harvesting designs proposed in this book.

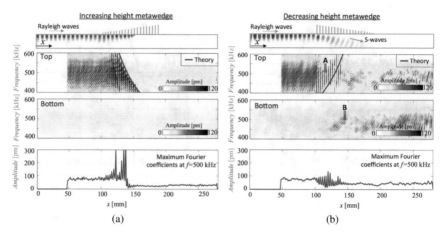

(a)   (b)

Fig. 2.22 a Space-frequency analysis of the experimental data (displacement) for the increasing metawedge. From top to bottom: the geometry, and field. The next two panels show the power spectra along the x-direction in the aluminium block measured at the centreline for the top and bottom surfaces. The white area in the surface plot is due to the transducer. The lowest plot represents the maximum value of the Fourier coefficients at 500 kHz for the scan located approximately at the centre of the top surface (y = 0). **b** Same as **a** but for the inverse metawedge. The blue line in the top scan indicates the theoretical prediction of the turning point at the surface. Note that points A and B are used to measure the refraction angle associated with conversion, see text for more details [58]

References

1. M. Carrara, M.R. Cacan, J. Toussaint, M.J. Leamy, M. Ruzzene, A. Erturk, Metamaterial-inspired structures and concepts for elastoacoustic wave energy harvesting. Smart Mater. Struct. (2013)
2. D. Torrent, J. Sánchez-Dehesa, Acoustic metamaterials for new two-dimensional sonic devices. New J. Phys. (2007)
3. O.R. Bilal, A. Foehr, C. Daraio, Bistable metamaterial for switching and cascading elastic vibrations. Proc. Natl. Acad. Sci. USA (2017)
4. M. Carrara, M.R. Cacan, M.J. Leamy, M. Ruzzene, A. Erturk, Dramatic enhancement of structure-borne wave energy harvesting using an elliptical acoustic mirror. Appl. Phys. Lett. (2012)
5. L.Y. Wu, L.W. Chen, C.M. Liu, Acoustic energy harvesting using resonant cavity of a sonic crystal. Appl. Phys. Lett. (2009)
6. L.Y. Wu, L.W. Chen, C.M. Liu, Acoustic pressure in cavity of variously sized two-dimensional sonic crystals with various filling fractions. Phys. Lett. Sect. A: Gen. At. Solid State Phys. (2009)
7. L.Y. Wu, L.W. Chen, C.M. Liu, Experimental investigation of the acoustic pressure in cavity of a two-dimensional sonic crystal. Phys. B: Condens. Matter (2009)
8. S. Qi, M. Oudich, Y. Li, B. Assouad, Acoustic energy harvesting based on a planar acoustic metamaterial. Appl. Phys. Lett. (2016)
9. W.C. Wang, L.Y. Wu, L.W. Chen, C.M. Liu, Acoustic energy harvesting by piezoelectric curved beams in the cavity of a sonic crystal. Smart Mater. Struct. (2010)
10. M. Oudich, Y. Li, Tunable sub-wavelength acoustic energy harvesting with a metamaterial plate. J. Phys. D: Appl. Phys. (2017)

11. A.C. To, W.K. Liu, G.B. Olson, T. Belytschko, W. Chen, M.S. Shephard, Y.W. Chung, R. Ghanem, P.W. Voorhees, D.N. Seidman, C. Wolverton, J.S. Chen, B. Moran, A.J. Freeman, R. Tian, X. Luo, E. Lautenschlager, A.D. Challoner, Materials integrity in microsystems: a framework for a petascale predictive-science-based multiscale modeling and simulation system. Comput. Mech. (2008)

12. S. Gonella, A.C. To, W.K. Liu, Interplay between phononic bandgaps and piezoelectric microstructures for energy harvesting. J. Mech. Phys. Solids (2009)

13. W.-P. Yang, L.-W. Chen, The tunable acoustic band gaps of two-dimensional phononic crystals with a dielectric elastomer cylindrical actuator. Smart Mater. Struct. (2008)

14. P.G. Martinsson, A.B. Movchan, Vibrations of lattice structures and phononic band gaps. Q. J. Mech. Appl. Math. (2003)

15. J.S. Jensen, Phononic band gaps and vibrations in one- and two-dimensional mass-spring structures. J. Sound Vib. (2003)

16. M. Hirsekorn, P.P. Delsanto, A.C. Leung, P. Matic, Elastic wave propagation in locally resonant sonic material: comparison between local interaction simulation approach and modal analysis. J. Appl. Phys. (2006)

17. Z. Chen, Y. Yang, Z. Lu, Y. Luo, Broadband characteristics of vibration energy harvesting using one-dimensional phononic piezoelectric cantilever beams. Phys. B: Condens. Matter (2013)

18. Y. Li, E. Baker, T. Reissman, C. Sun, W.K. Liu, Design of mechanical metamaterials for simultaneous vibration isolation and energy harvesting. Appl. Phys. Lett. (2017)

19. K. Mikoshiba, J.M. Manimala, C.T. Sun, Energy harvesting using an array of multifunctional resonators. J. Intell. Mater. Syst. Struct. (2013)

20. J. Li, X. Zhou, G. Huang, G. Hu, Acoustic metamaterials capable of both sound insulation and energy harvesting. Smart Mater. Struct. (2016)

21. X. Wang, J. Xu, J. Ding, C. Zhao, Z. Huang, A compact and low-frequency acoustic energy harvester using layered acoustic metamaterials. Smart Mater. Struct. (2019)

22. X. Zhang, H. Zhang, Z. Chen, G. Wang, Simultaneous realization of large sound insulation and efficient energy harvesting with acoustic metamaterial. Smart Mater. Struct. (2018)

23. H. Nguyen, R. Zhu, J.K. Chen, S.L. Tracy, G.L. Huang, Analytical coupled modeling of a magneto-based acoustic metamaterial harvester. Smart Mater. Struct. (2018)

24. K.H. Sun, J.E. Kim, J. Kim, K. Song, Sound energy harvesting using a doubly coiled-up acoustic metamaterial cavity. Smart Mater. Struct. (2017)

25. S. Qi, Y. Li, B. Assouar, Acoustic focusing and energy confinement based on multilateral metasurfaces. Phys. Rev. Appl. (2017)

26. R. Ahmed, D. Madisetti, S. Banerjee, A sub-wavelength scale acoustoelastic sonic crystal for harvesting energies at very low frequencies using controlled geometric configurations. J. Intell. Mater. Syst. Struct. (2017)

27. R.U. Ahmed, S. Banerjee, Low frequency energy scavenging using sub-wave length scale acousto-elastic metamaterial. AIP Adv. (2014)

28. C. Sugino, V. Guillot, A. Erturk, Multifunctional energy harvesting locally resonant metastructures, in *ASME 2017 Conference on Smart Materials, Adaptive Structures and Intelligent Systems, SMASIS 2017* (2017)

29. C. Sugino, A. Erturk, Analysis of multifunctional piezoelectric metastructures for low-frequency bandgap formation and energy harvesting. J. Phys. D: Appl. Phys. (2018)

30. C. Sugino, S. Leadenham, M. Ruzzene, A. Erturk, On the mechanism of bandgap formation in locally resonant finite elastic metamaterials. J. Appl. Phys. (2016)

31. C. Sugino, Y. Xia, S. Leadenham, M. Ruzzene, A. Erturk, A general theory for bandgap estimation in locally resonant metastructures. J. Sound Vib. (2017)

32. S. Tol, F.L. Degertekin, A. Erturk, Gradient-index phononic crystal lens-based enhancement of elastic wave energy harvesting. Appl. Phys. Lett. (2016)

33. S.C.S. Lin, T.J. Huang, J.H. Sun, T.T. Wu, Gradient-index phononic crystals. Phys. Rev. B Condens. Matter Mater. Phys. (2009)

34. Y. Jin, D. Torrent, Y. Pennec, Y. Pan, B. Djafari-Rouhani, Simultaneous control of the S0 and A0 Lamb modes by graded phononic crystal plates. J. Appl. Phys. (2015)

35. X. Yan, R. Zhu, G. Huang, F.G. Yuan, Focusing guided waves using surface bonded elastic metamaterials. Appl. Phys. Lett. (2013)
36. J. Zhao, R. Marchal, B. Bonello, O. Boyko, Efficient focalization of antisymmetric Lamb waves in gradient-index phononic crystal plates. Appl. Phys. Lett. (2012)
37. T.-T. Wu, M.-J. Chiou, Y.-C. Lin, T. Ono, Design and fabrication of a gradient-index phononic quartz plate lens, in *Photonic and Phononic Properties of Engineered Nanostructures IV* (2014)
38. A. Climente, D. Torrent, J. Sánchez-Dehesa, Gradient index lenses for flexural waves based on thickness variations. Appl. Phys. Lett. (2014)
39. A. Zareei, A. Darabi, M.J. Leamy, M.R. Alam, Continuous profile flexural GRIN lens: focusing and harvesting flexural waves. Appl. Phys. Lett. (2018)
40. S. Tol, F.L. Degertekin, A. Erturk, Phononic crystal Luneburg lens for omnidirectional elastic wave focusing and energy harvesting. Appl. Phys. Lett. (2017)
41. L. Zhao, S.C. Conlon, F. Semperlotti, Broadband energy harvesting using acoustic black hole structural tailoring. Smart Mater. Struct. (2014)
42. L. Zhao, S.C. Conlon, F. Semperlotti, An experimental study of vibration based energy harvesting in dynamically tailored structures with embedded acoustic black holes. Smart Mater. Struct. (2015)
43. M.A. Mironov, Propagation of a flexural wave in a plate whose thickness decreases smoothly to zero in a finite interval. Sov. Phys. Acoust.-USSR (1988)
44. M.A. Mironov, V.V. Pislyakov, One-dimensional acoustic waves in retarding structures with propagation velocity tending to zero. Acoust. Phys. (2002)
45. V.V. Krylov, F.J.B.S. Tilman, Acoustic 'black holes' for flexural waves as effective vibration dampers. J. Sound Vib. (2004)
46. V.V. Krylov, New type of vibration dampers utilising the effect of acoustic 'black holes'. Acta Acust. united Acust. (2004)
47. V.V. Krylov, R.E.T.B. Winward, Experimental investigation of the acoustic black hole effect for flexural waves in tapered plates. J. Sound Vib. (2007)
48. D.J. O'Boy, V.V. Krylov, V. Kralovic, Damping of flexural vibrations in rectangular plates using the acoustic black hole effect. J. Sound Vib. (2010)
49. V.B. Georgiev, J. Cuenca, F. Gautier, L. Simon, V.V. Krylov, Damping of structural vibrations in beams and elliptical plates using the acoustic black hole effect. J. Sound Vib. (2011)
50. K.L. Tsakmakidis, A.D. Boardman, O. Hess, 'Trapped rainbow' storage of light in metamaterials. Nature (2007)
51. J. Zhu, Y. Chen, X. Zhu, F.J. Garcia-Vidal, X. Yin, W. Zhang, X. Zhang, Acoustic rainbow trapping. Sci. Rep. (2013)
52. X. Ni, Y. Wu, Z.G. Chen, L.Y. Zheng, Y.L. Xu, P. Nayar, X.P. Liu, M.H. Lu, Y.F. Chen, Acoustic rainbow trapping by coiling up space. Sci. Rep. (2014)
53. Y. Chen, H. Liu, M. Reilly, H. Bae, M. Yu, Enhanced acoustic sensing through wave compression and pressure amplification in anisotropic metamaterials. Nat. Commun. (2014)
54. V. Romero-García, R. Picó, A. Cebrecos, V.J. Sánchez-Morcillo, K. Staliunas, Enhancement of sound in chirped sonic crystals. Appl. Phys. Lett. (2013)
55. A. Cebrecos, R. Picó, V.J. Sánchez-Morcillo, K. Staliunas, V. Romero-García, L.M. Garcia-Raffi, Enhancement of sound by soft reflections in exponentially chirped crystals. AIP Adv. (2014)
56. D. Cardella, P. Celli, S. Gonella, Manipulating waves by distilling frequencies: a tunable shunt-enabled rainbow trap. Smart Mater. Struct. (2016)
57. A. Colombi, D. Colquitt, P. Roux, S. Guenneau, R.V. Craster, A seismic metamaterial: the resonant metawedge. Sci. Rep. (2016)
58. A. Colombi, V. Ageeva, R.J. Smith, A. Clare, R. Patel, M. Clark, D. Colquitt, P. Roux, S. Guenneau, R.V. Craster, Enhanced sensing and conversion of ultrasonic Rayleigh waves by elastic metasurfaces. Sci. Rep. (2017)
59. Z. Tian, L. Yu, Rainbow trapping of ultrasonic guided waves in chirped phononic crystal plates. Sci. Rep. (2017)

60. A.E. Miroshnichenko, S. Flach, Y.S. Kivshar, Fano resonances in nanoscale structures. Rev. Mod. Phys. (2010)
61. A. Weinmann, L.D. Landau, E.M. Lifshitz, J.B. Sykes, J.S. Bell, Quantum mechanics (Non-relativistic theory). Math. Gaz. (1959)
62. N.C. Perkins, C.D. Mote, Comments on curve veering in eigenvalue problems. J. Sound Vib. (1986)

Chapter 3
One-Dimensional Inhomogeneous Media

Abstract This chapter analyses one-dimensional inhomogeneous media through lumped models. Basic concepts on wave propagation in homogeneous and inhomogeneous media are introduced for both periodic and aperiodic structures. An interpretation of the most common linear wave propagation mechanisms is provided, alternating heuristic approaches, with purely analytical, numerical and experimental results. The design of an aperiodic inhomogeneous structure for vibration isolation is also presented.

3.1 The Propagation of Mechanical Waves

A *wave* can be defined as the propagation of a disturbance with oscillations about a stable equilibrium configuration. As the disturbance propagates, it carries along amounts of energy that can be transmitted over considerable distances. A mechanical wave is a local strain that propagates in a deformable body from particle to particle, by creating local stresses. Specifically, to create a wave, two opposed forces that simultaneously counteract and restore equilibrium are required. This is done by the *inertia* and *elastic* forces which correspond, energetically speaking, to the *kinetic* and *potential* energies, as shown in Fig. 3.1a. If we lose the elastic force, it is not a wave, but motion of mass. If the wave is defined by nodes and anti-nodes (see Fig. 3.1b), i.e. with a fixed peaks amplitude profile in space, it is called *stationary*; contrary, it is called *travelling* or *propagating*. A particular wave is the *plane wave*, that is a wave with constant amplitude for any plane perpendicular to its direction of propagation. It is worth to notice that the direction of *propagation* does not necessarily coincide with the direction of oscillation of the particles: if equal, the wave is called *longitudinal*,

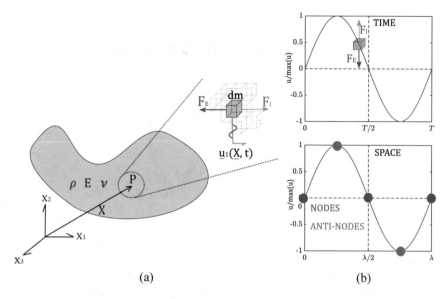

Fig. 3.1 a Schematic of a mechanical wave in a homogeneous continuum: if a perturbation u_1 is applied, each infinitesimal continuum volume oscillates under elastic and inertia forces. **b** Period T and wavelength λ of a wave in time and space domain respectively, and definition of nodes and anti-nodes for a standing wave

otherwise *transverse*. We introduce the basic scalar nomenclature adopted in the text for the physics of waves:

T	period	[s]
f	frequency	[1/s]
ω	angular frequency	[rad/s]
λ	wavelength	[m]
κ	(angular) wavenumber	[rad/m]
c	celerity, or wave velocity	[m/s]

An elastic, homogeneous and isotropic continuum, is defined by three parameters: density ρ, Young (or Elastic) modulus E, and Poisson ratio ν. It is reasonable to assume that, in this medium, the wave velocity is $c = c(\rho, E, \nu)$.

By adopting the form $c = \gamma(\nu)\rho^\alpha E^\beta$, due to dimensional considerations we have:

$$\begin{cases} [m] \ : 1 = -3\alpha - \beta, \\ [s] \ : -1 = -2\beta \to \beta = 1/2, \\ [kg] \ : 0 = \alpha + \beta \to \alpha = -1/2, \end{cases} \tag{3.1}$$

and then: $c = \gamma(\nu)\sqrt{E/\rho}$, which means that increasing the elastic parameters or decreasing the inertial one, the wave velocity increases. The constant γ is a function of the Poisson's ratio, and it is equal to one for the case of a beam (see the next section).

3.2 One-Dimensional Periodic Structures

As implicitly stated by their name, periodic structures are systems composed of a periodic repetition in space of an elementary component, usually called *unit* or *primitive* cell. This constitutive element can be univocally identified only if a periodic spatial variation of the medium properties exists. In elasticity, this can be achieved through a periodic variation of the inertial or elastic parameters (for e.g. mass density or elastic constants). We start considering the simplest case of the 1D wave equation (d'Alambert 1747) for longitudinal waves in a homogeneous beam. Considering an infinitesimal element of length dx (Fig. 3.2), the horizontal equilibrium equation reads:

$$- N(x,t) + N(x,t) + \frac{\partial N(x,t)}{\partial x} dx - \rho(x,t) A(x,t) dx \frac{\partial^2 u(x,t)}{\partial t^2} = 0, \quad (3.2)$$

with $N(x,t)$ the axial force, $\rho(x,t)$ the mass density and $A(x,t)$ the rod's cross section.

Introducing the Hooke's law ($N = EA\varepsilon$) and the compatibility equation ($\varepsilon = \partial u/\partial x$), we get:

$$\frac{\partial}{\partial x} \left[E(x,t) A(x,t) \frac{\partial u(x,t)}{\partial x} \right] dx - \rho(x,t) A(x,t) dx \frac{\partial^2 u(x,t)}{\partial t^2} = 0, \quad (3.3)$$

with $E(x,t)$ a space-time variant Young's modulus. Assuming for simplicity $E(x,t) = E_0$, $A(x,t) = A_0$ and $\rho(x,t) = \rho_0$, we get the well known equation:

$$\frac{\partial^2 u(x,t)}{\partial t^2} = c^2 \frac{\partial^2 u(x,t)}{\partial x^2}, \quad (3.4)$$

Fig. 3.2 Longitudinal waves in a homogeneous elastic beam. The wave equation is obtained imposing the horizontal equilibrium of an infinitesimal element of length dx under elastic (axial) and inertia forces

with $c = \sqrt{E_0/\rho_0}$ (as heuristically previously defined for $\gamma = 1$). Assuming now the simplest case of a *scalar harmonic plane wave*:

$$u(x, t) = \tilde{u}e^{i(\pm\kappa x \pm \omega t)}, \tag{3.5}$$

(where $\pm$ is an arbitrary choice), and introducing Eq. (3.5) into Eq. (3.4), the following is obtained (for positive angular frequency):

$$\omega = c\kappa. \tag{3.6}$$

We notice that Eq. (3.5) represents a vector in the complex plane with magnitude $\tilde{u}$, rotating with an angle ωt for a given position, or κx for a given instant of time. The projection of this vector on the real and imaginary axes defines two sinusoidal functions in quadrature (i.e. orthogonal to each other).

An equation like Eq. (3.6) between the angular frequency and the wavenumber is called *dispersion relation*. From the definition of the angular frequency ($\omega = 2\pi f$) and the wavenumber ($\kappa = 2\pi/\lambda$), one simply gets:

$$c = \lambda f. \tag{3.7}$$

The *phase* velocity, i.e. the rate at which the phase of the wave propagates in space, is then defined from Eq. (3.6) as $c_{ph} = \omega/\kappa$. On the other hand, the *group* velocity, i.e. the velocity with which the overall envelope shape of the wave's amplitudes propagates through space (wave packet velocity), is found by setting, due to linearity, a constant spatiotemporal phase variation:

$$\Delta\varphi = \Delta\kappa x - \Delta\omega t = constant, \tag{3.8}$$

which, when differentiated yields:

$$\frac{dx}{dt} = \frac{\Delta\omega}{\Delta\kappa}. \tag{3.9}$$

In the limit $\Delta\kappa \to 0$, we obtain the group velocity definition as:

$$c_g = \frac{\partial\omega}{\partial\kappa}, \tag{3.10}$$

which is just the same as the velocity of energy transport of a monochromatic wave [1]. If the dispersion relation is linear, c_{ph} and c_g are equal, and the medium is then called *non-dispersive*. There are several ways of expressing the group velocity:

$$c_g = \frac{\partial\omega}{\partial\kappa} = \frac{\partial}{\partial\kappa}[c_{ph}\kappa] = c_{ph} + \kappa\frac{\partial c_{ph}}{\partial\kappa}, \tag{3.11}$$

or, alternatively:

Fig. 3.3 Transverse waves in a homogeneous elastic beam. The wave equation is obtained imposing the vertical equilibrium of an infinitesimal element of length dx under elastic (shear) and inertia forces

$$\frac{1}{c_g} = \frac{\partial \kappa}{\partial \omega} = \frac{\partial}{\partial \omega}\left[\frac{\partial \omega c_{ph} - \omega \partial c_{ph}}{c_{ph}^2}\right] = \frac{1}{c_{ph}} - \frac{\omega}{c_{ph}^2}\frac{\partial c_{ph}}{\partial \omega}. \qquad (3.12)$$

In general, $\partial c/\partial \kappa < 0$ and thus $c_g < c_{ph}$. If $\partial c/\partial \kappa < 0$, it is called *normal dispersion*, otherwise if $\partial c/\partial \kappa > 0$ and $c_g > c_{ph}$, *anomalous dispersion*. It can be noticed that if the medium is non-dispersive, for a given frequency of the propagating wave, the wavelength is constant, and without any spatial variation.

However, it is important to outline that the dispersion properties are strongly related to the nature of the propagating wave, and not only to the medium. For this reason, we consider now the same beam, but with transverse waves. Considering, as before, an infinitesimal element of length dx (Fig. 3.3), the vertical equilibrium equation reads:

$$T(x, t) - T(x, t) - \frac{\partial T(x, t)}{\partial x}dx - \rho(x, t)A(x, t)dx\frac{\partial^2 v(x, t)}{\partial t^2} = 0, \qquad (3.13)$$

with $T(x, t)$ the shear force, $\rho(x, t)$ the mass density and $A(x, t)$ the rod cross section.

Introducing the equilibrium relation between bending moment and shear force ($T = \partial M/\partial x$), and the constitutive law ($M = EI[\partial^2 v/\partial x^2]$), we get:

$$\frac{\partial}{\partial x}\left\{\frac{\partial}{\partial x}\left[E(x, t)I(x, t)\frac{\partial^2 v(x, t)}{\partial x^2}\right]\right\}dx + \rho(x, t)A(x, t)dx\frac{\partial^2 v(x, t)}{\partial t^2} = 0, \qquad (3.14)$$

with $E(x, t)$ and $I(x, t)$ a space-time variant Young's modulus and second moment of inertia respectively. Assuming again, for simplicity, $E(x, t) = E_0$, $A(x, t) = A_0$, $\rho(x, t) = \rho_0$ and $I(x, t) = I_0$, we obtain:

$$E_0 I_0 \frac{\partial^4 v(x, t)}{\partial x^4} + \rho_0 A_0 \frac{\partial^2 v(x, t)}{\partial t^2} = 0. \qquad (3.15)$$

Introducing Eq. (3.5) into Eq. (3.15), the following dispersion relation is obtained (for positive angular frequency):

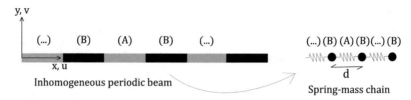

Fig. 3.4 Inhomogeneous beam with a spatial periodic modulation of the elastic properties, and corresponding monoatomic spring-mass chain idealization. Since material (A) is light and soft, it is modeled using springs, while material (B), heavy and stiff, using lumped masses

$$\omega = \sqrt{\frac{E_0 I_0}{\rho_0 A_0} \kappa^2}. \tag{3.16}$$

With respect to the previous case, the dispersion relation is now nonlinear, and the wave propagation is then *dispersive*. As for the previous case, from the definition of the phase and group velocity we get:

$$c_p = \omega/\kappa = \sqrt{\frac{E_0 I_0}{\rho_0 A_0}} \kappa, \tag{3.17}$$

$$c_g = \frac{\partial \omega}{\partial \kappa} = 2 \sqrt{\frac{E_0 I_0}{\rho_0 A_0}} \kappa. \tag{3.18}$$

It is important to notice that a simple way to obtain a *dispersive* behaviour (also for longitudinal waves), is to modify the elastic or inertial parameters, for e.g. through a spatial or temporal modulation of the medium properties.

Let us assume now the simple case of a medium with a spatial periodic modulation of the elastic properties. This can be achieved by considering two different materials (light-soft, heavy-stiff), or varying the cross section of the beam (Fig. 3.4).

This kind of system is the simplest periodic structure, whose *elementary* cell is its smallest part periodically repeated along space. The simplest model is a spring-mass chain, firstly analysed in Newton *Principia* (1686), and then by John Bernoulli and son Daniel (1727). While the original formulation (Newton), was obtained writing the dynamic equilibrium of the ith mass inside the chain, we propose here another approach starting from the continuous Eq. (3.3). Going from continuous to discrete $u(x, t) \mapsto u(x_n, t) = u_n(t)$, we get:

$$E_0 A_0 \frac{u_{n-1}(t) - 2u_n(t) + u_{n+1}(t)}{d^2} d - \rho_0 A_0 d \ddot{u}_n(t) = 0, \tag{3.19}$$

where d is the distance between two adjacent masses. Rearranging Eq. (3.19) and defining the stiffness $k_S = E_0 A_0/d$ and mass $m = \rho_0 A_0 d$, we get:

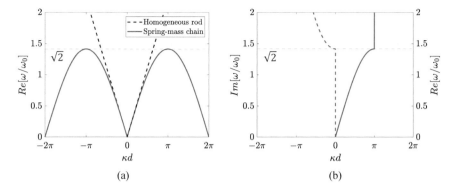

Fig. 3.5 a Dispersion relation for longitudinal waves in a homogeneous (dashed black) and inhomogeneous (red) beam, modeled through a monoatomic spring-mass chain with equivalent elastic constants. **b** Real and imaginary part of the dispersion relation

$$m\ddot{u}_n(t) + 2k_S u_n(t) - k_S(u_{n-1} + u_{n+1}) = 0. \tag{3.20}$$

Substituting the plane wave harmonic solution of Eq. (3.5), as for the previous cases, the following dispersion relation is obtained:

$$-\omega^2 m + 2k_S - k_S[e^{-i\kappa d} + e^{i\kappa d}] = 0, \tag{3.21}$$

from which (positive angular frequency):

$$\omega = \sqrt{2\frac{k_S}{m}[1 - \cos(\kappa d)]}. \tag{3.22}$$

The dispersion relation reported in Eq. (3.22) is symmetric about $\kappa = 0$ and periodic with period 2π, as shown in Fig. 3.5a.

The behaviour of the relation in the fundamental range $\kappa \in [-\pi/d, \ \pi/d]$, therefore fully describes the characteristics of harmonic plane waves in terms of angular frequency and wavenumber. This range is known as *First Brillouin Zone* (FBZ) [2]. The symmetry of the dispersion relation provides the opportunity to fully characterize the dispersion properties considering half of the FBZ, i.e. $\kappa \in [0, \ \pi/d]$, which is known as *Irreducible Brillouin Zone* (IBZ).

It is important to notice that Eq. (3.22) has a top bound given by the frequency $\omega_{BG} = \sqrt{2}\sqrt{2k_S/m} = \sqrt{2}\omega_0$, with ω_0 the natural angular frequency of the unit cell. Therefore, for $\omega > \sqrt{2}\omega_0$, there are no corresponding real-valued wavenumbers, and harmonic plane waves do not propagate. The range of $\omega > \sqrt{2}\omega_0$ corresponds to an attenuation range, also denoted in literature as *stop band* or, when occurs over a limited frequency range, *band gap*. The amount of attenuation can be quantified by solving the dispersion relation at a given frequency in terms of the wavenumber, which allows the estimation of both the real and imaginary parts as a function of

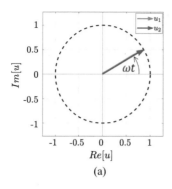 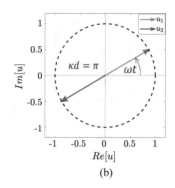

(a)   (b)

Fig. 3.6 a In-phase and **b** out-of-phase response of adjacent point masses. Band gap opening at the edge of the first BZ is due to out-of-phase response, i.e. wave reflection

frequency (see Fig. 3.5b):

$$\kappa d = \arccos\left(1 - \frac{m}{2k_S}\omega^2\right). \tag{3.23}$$

From the definition of the phase and group velocity, we get:

$$c_p = \omega/\kappa = \frac{1}{\kappa}\sqrt{2\frac{k_S}{m}[1 - \cos(\kappa d)]}, \tag{3.24}$$

$$c_g = \frac{\partial \omega}{\partial \kappa} = \frac{\frac{k_S d}{m}\sin(\kappa d)}{\sqrt{2\frac{k_S}{m}[1 - \cos(\kappa d)]}}, \tag{3.25}$$

from which:

$$c = \lim_{\kappa \to 0} c_p(\kappa d) = \lim_{\kappa \to 0} c_g(\kappa d) = \sqrt{\frac{k_S}{m}}d = \sqrt{\frac{E_0}{\rho_0}}. \tag{3.26}$$

Specifically, the group velocity is zero for $\kappa d = \pi$, i.e. at the edge of the first BZ. This condition is associated to a π shift of the wave, i.e. wave reflection (Fig. 3.6). By simply writing the plane wave solution reported in Eq. (3.5) between two adjacent point masses we get:

$$\tilde{u}_1 e^{i[\kappa x_1 + \omega t]} = \tilde{u}_2 e^{i[\kappa(x_1 + d) + \omega t]} \Rightarrow \tilde{u}_1 = \tilde{u}_2 e^{i\kappa d}. \tag{3.27}$$

For $\kappa d = \pi$ we obtain an out-of-phase ($\tilde{u}_1 = -\tilde{u}_2$) response. This is a simple demonstration that the zero group velocity mode achieved at the edge of the first BZ is associated to wave *reflection*.

We point out that the predicted band gap is due to the discrete nature of the system. Considering Fig. 3.7a, it can be noticed, due to Nyquist frequency, that:

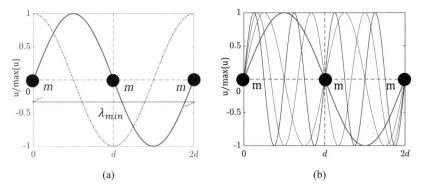

Fig. 3.7 **a** Maximum wavelength for a spring-mass chain and **b** aliasing for waves with $\lambda < \lambda_{min}$

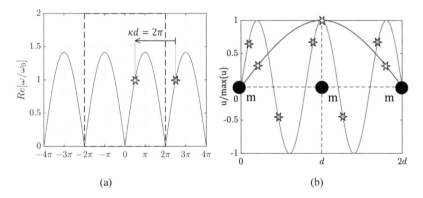

Fig. 3.8 **a** Dispersion relation inside and outside the first BZ for a spring-mass chain. A wavevector outside the first BZ can be introduced inside through a shift of $2\pi/d$. **b** Equivalent solutions for the mass displacement outside (red curve and grey star) and inside (grey curve and yellow star) the first BZ

$$\lambda_{min} = 2d \quad \rightarrow \quad \kappa_{max} = \frac{2\pi}{\lambda_{min}} = \frac{\pi}{d}, \tag{3.28}$$

which is a simple demonstration of the 1D IBZ. In addition, related to this wavelength, we get the maximum resonance as $\omega_{max} = \sqrt{2k_S/m}$, since the two external masses move in-phase while the internal one out-of -phase (see the grey curve in Fig. 3.7a). It is important to notice that, for $\lambda < \lambda_{min}$, we have aliasing, i.e. the different signals sampled on the masses become indistinguishable, as shown in Fig. 3.7b.

However, it is possible to obtain an equivalent solution for the mass displacements by applying a shift of $\kappa d = 2\pi$, i.e. $\kappa = 2\pi/d$, as shown by the dispersion in Fig. 3.8a and the wave amplitude profile in Fig. 3.8b. This concept is at the base of a phenomenon known in physics as *Umklapp*, that will be discussed later.

The direct consequence of this concept is that the most significant wavevectors (i.e. the vectorial counterpart of the wavenumber) for a 2D problem with a square unit cell will be $\kappa_x = (\pi/d)i$, $\kappa_y = (\pi/d)j$ and $\kappa_{xy} = (\pi/d)i + (\pi/d)j$ (with i and j orthogonal versors). Specifically, we will expect the maximum wavenumber (i.e. lower wavelength) as:

$$\kappa_{max} = \sqrt{\kappa_{x,max}^2 + \kappa_{y,max}^2} = \sqrt{2}\frac{\pi}{d} \tag{3.29}$$

This path $\Gamma(0, 0) \rightarrow X(\pi/d, 0) \rightarrow M(\pi/d, \pi/d) \rightarrow \Gamma(0, 0)$ defines the 2D IBZ for a square unit cell.

Band gap interpretation. We provide now an interpretation of the concept of band gap, using elementary dynamics and energy considerations.

As stated in the previous section, a band gap can be defined as the frequency range for which the wave propagation is forbidden. From a different perspective, in this frequency range, only *evanescent* waves exist (see Fig. 3.5b). This means that the wave do not propagate in the medium, and the energy is spatially concentrated in the vicinity of the source. A hallmark of an evanescent wave is that there is no energy flow in the considered region, with a *Poynting vector* (i.e. the directional energy flux) equal to zero. We firstly emphasize that a stop band like the one in Eq. (3.22), is impossible to be obtained in real media. As a matter of fact, this band gap is due to the discrete nature of the system. While for real 3D continuous media the wave is supported by an infinite number of *propagating modes*, this is not true for discrete systems. Even if the spring-mass chain is infinite, the number of propagating modes is finite due to the discretisation process.

We show now that all these concepts can be elementary deduced isolating the single cell of the spring-mass chain [3]. For the single cell, composed of two masses $m/2$ connected with a spring of stiffness k_S and subject to a generic force $F(t)$, the equations of motions are:

$$\begin{cases} \frac{m}{2}\ddot{u}_1(t) + k_S[u_1(t) - u_2(t)] = F(t), \\ \frac{m}{2}\ddot{u}_2(t) + k_S[u_2(t) - u_1(t)] = 0. \end{cases} \tag{3.30}$$

Applying a harmonic force $F(t) = \tilde{F}e^{i\omega t}$, in steady state regime the response also becomes harmonic in the form of $u_i(t) = \tilde{u}_i e^{i\omega t}$. Computing from the second of Eq. (3.30), the ratio $|\tilde{u}_2|/|\tilde{u}_1|$, the following is obtained:

$$\left|\frac{\tilde{u}_2}{\tilde{u}_1}\right| = \left|\frac{k_S}{-\frac{m}{2}\omega^2 + k_S}\right| = \left|\frac{1}{1 - \beta^2}\right| = N(\beta), \tag{3.31}$$

where $\beta = \omega/\omega_0$. $N(\beta)$ represents the well known magnification factor for an undamped Single Degree Of Freedom (SDOF) oscillator under a harmonic force. It is clear that the structure starts to provide attenuation if $N(\beta) < 1$, which corresponds to $\beta > \sqrt{2}$ (Fig. 3.9a). An attenuation regime is therefore obtained after the angular frequency:

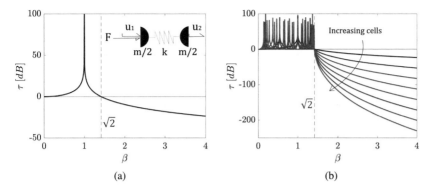

Fig. 3.9 **a** Transmission function for a single cell made of two masses connected with a spring (see inset) and **b** for different number of cells. Increasing the number of cells, the attenuation increases, and the transmission function takes always the value of zero for $\beta = \sqrt{2}$

$$\omega_{att.} = \sqrt{2}\beta = \sqrt{2}\sqrt{\frac{2k_S}{m}} = 2\sqrt{\frac{k_S}{m}}, \tag{3.32}$$

which is coherent with the dispersion relation for a spring-mass chain in Eq. (3.22). The addition of cells does not alter the band gap, but creates other modes that are always located at a position below $\beta = \sqrt{2}$ (Fig. 3.9b). Furthermore, whichever is the number of cells, the transmission function $\tau = 20log_{10}[N(\beta)]$ takes the value of zero for $\beta = \sqrt{2}$ (Fig. 3.9b).

Due to the analogy between the dynamics of the infinite chain and the single cell, we draw some energy considerations and phase arguments looking at the unit cell only. Solving the dynamic equilibrium equations in Eq. (3.30), in steady state harmonic regime we get:

$$\tilde{u}_1(\beta) = \frac{\tilde{F}}{k_S} \frac{1 - \beta^2}{\beta^2(\beta^2 - 2)}, \tag{3.33}$$

$$\tilde{u}_2(\beta) = \frac{\tilde{F}}{k_S} \frac{1}{\beta^2(\beta^2 - 2)}. \tag{3.34}$$

As expected, the condition $\beta = \sqrt{2}$ is related to a resonance with an out-of-phase response of the two point masses (Fig. 3.10a). After this value, the motion of the first mass (input) increases, while the motion of the second (output) decreases (Fig. 3.10a). It is then possible to write the strain energy, kinetic energy and external work amplitudes as:

$$\tilde{U}(\beta) = \frac{1}{2}k_S[\tilde{u}_1 - \tilde{u}_2]^2 = \frac{1}{2}\frac{\tilde{F}^2}{k_S}\left[\frac{1}{(\beta^2 - 2)^2}\right], \tag{3.35}$$

$$\tilde{T}(\beta) = \frac{1}{2}\frac{m}{2}[-\omega^2(\tilde{u}_1^2 + \tilde{u}_2^2)] = -\frac{1}{2}\beta^2\frac{\tilde{F}^2}{k_S}\left[\frac{(1 - \beta^2)^2 + 1}{\beta^4(\beta^2 - 2)^2}\right], \tag{3.36}$$

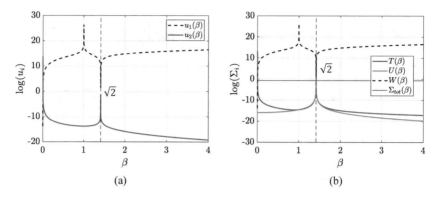

Fig. 3.10 **a** Displacement field of the first and second point mass of a single cell in logarithmic scale. **b** Energy balance (logarithmic scale) for different values of β. After $\beta = \sqrt{2}$ the external work increases, while the kinetic and strain energy decrease monotonically

$$\tilde{W}(\beta) = \tilde{F}\tilde{u}_1 = \frac{\tilde{F}^2}{k_S}\left[\frac{1 - \beta^2}{\beta^2(\beta^2 - 2)}\right]. \tag{3.37}$$

It can be noticed that for $\beta > \sqrt{2}$, the external work increases monotonically, while the kinetic and strain energy decrease monotonically (Fig. 3.10b). This corroborates that the energy remains confined in the input region for a band gap at the edge of the first BZ. The energy conservation principle is always satisfied since the total energy Σ_{tot} is constant (Fig. 3.10b).

We now analyse the same problem through Argand diagrams, i.e. by looking at the elastic, inertia and external forces in the complex plane (Fig. 3.11). The equations of the elastic and inertia force for the first mass are respectively:

$$f_{EL}^{(1)}(\beta, t) = -k_S(u_1 - u_2) = \tilde{F}\frac{\beta^2}{\beta^2(\beta^2 - 2)}e^{i\omega t}, \tag{3.38}$$

$$f_I^{(1)}(\beta, t) = -m\ddot{u}_1 = \tilde{F}\beta^2\frac{1 - \beta^2}{\beta^2(\beta^2 - 2)}e^{i\omega t}, \tag{3.39}$$

while, for the second mass:

$$f_{EL}^{(2)}(\beta, t) = -k_S(u_2 - u_1) = -\tilde{F}\frac{\beta^2}{\beta^2(\beta^2 - 2)}e^{i\omega t}, \tag{3.40}$$

$$f_I^{(2)}(\beta, t) = -m\ddot{u}_2 = \tilde{F}\beta^2\frac{1}{\beta^2(\beta^2 - 2)}e^{i\omega t}. \tag{3.41}$$

It is possible to define three critical conditions for the first mass, and two for the second. Specifically, for the first mass:

- $\beta < 1$ the external force is out-of-phase with respect to both the elastic and inertia forces;
- $1 \leq \beta \leq \sqrt{2}$ the external and inertia forces are both out-of-phase with respect to the elastic force;
- $\beta > \sqrt{2}$ the external and elastic force are both out-of-phase with respect to the inertia force;

while, for the second mass, the elastic and inertia forces are always out-of-phase but with different orientations for the conditions:

- $\beta \leq \sqrt{2}$;
- $\beta > \sqrt{2}$.

It is important to notice that for $\beta > \sqrt{2}$, i.e. inside the vibration attenuation range, the contribution of the elastic force for the first mass goes to zero. This means that the first mass is governed by an *inertial* behaviour, with an external force equilibrated almost only by the inertia force. For the second mass instead, the behaviour is *elastic*, i.e. the elastic force counteracts the motion of the mass.

As previously explained, an elastic inhomogeneous medium with a periodic structure is usually called phononic crystal. In recent years, several phononic crystals designs have been proposed [4–7], especially for vibration isolation purposes [8–12]. Most of these works focus on optimization procedures in order to obtain wide band gaps [5, 13–16], usually quantified through a gap-to-mid gap ratio, i.e. the ratio between the gap width and the central frequency. However, this trend is slowing vanishing since the gap characteristics are inherently linked to some critical parameters, mainly the high impedance mismatch of the crystal, which gives low structural performances in terms of stiffness and strength [8–12].

3.3 One-Dimensional Aperiodic Structures

In the setting of elasticity, aperiodic structures are inhomogeneous media defined by a random distribution of the elastic parameters or cells. In order to understand their behaviour, we start with an enriched periodic chain with two types of punctual masses, and then move to full aperiodic systems.

We consider a beam made of three different materials (one light-soft and two heavy-stiff), or with a cross section variation that can be represented as a periodic spring-mass chain with two types of masses m_1 and m_2 (Fig. 3.12).

Defining as $u_n(t)$ the displacement of the mass m_1, and $v_n(t)$ the one of m_2, we get the following equilibrium equations:

$$\begin{cases} m_1 \ddot{u}_n + k_S(u_n - v_{n-1}) + k_S(u_n - v_n) = 0, \\ m_2 \ddot{v}_n + k_S(v_n - u_n) + k_S(v_n - u_{n+1}) = 0. \end{cases} \tag{3.42}$$

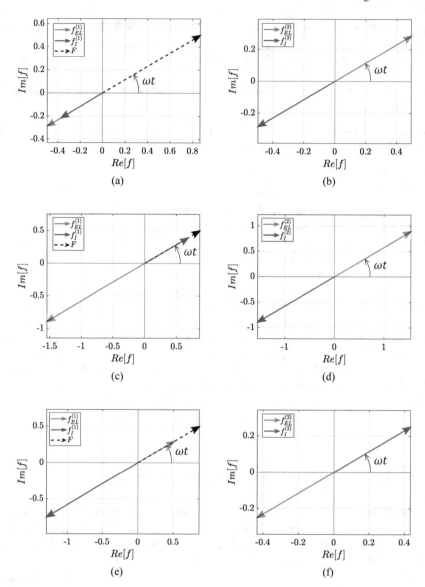

Fig. 3.11 Argand diagrams for a spring-mass unit cell for different values of β. First and second mass for $\beta = 0.5$ (**a–b**), $\beta = 1.2$ (**c–d**) and $\beta = 2$ (**e–f**) respectively

Substituting the plane wave harmonic solution reported in Eq. (3.5), and imposing, due to periodicity, equal amplitudes $\tilde{u}_n = \tilde{u}_{n+1}$ and $\tilde{v}_n = \tilde{v}_{n-1}$, the following linear system is obtained:

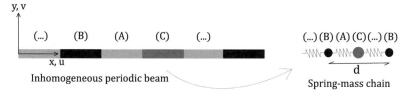

Fig. 3.12 Inhomogeneous beam with spatial periodic modulation of the elastic properties and diatomic spring-mass chain idealization. The beam is composed of a light-soft material (A) and two heavy-stiff materials (B) and (C)

$$\begin{bmatrix} (-\omega^2 m_1 + 2k_S) & -k_S(e^{i\kappa\frac{d}{2}} + e^{-i\kappa\frac{d}{2}}) \\ -k_S(e^{i\kappa\frac{d}{2}} + e^{-i\kappa\frac{d}{2}}) & (-\omega^2 m_2 + 2k_S) \end{bmatrix} \begin{pmatrix} \tilde{u}_n \\ \tilde{v}_n \end{pmatrix} = \mathbf{0}. \tag{3.43}$$

Nontrivial solutions are then obtained only if the determinant vanishes, leading to the following dispersion relation (for positive values):

$$\omega = \sqrt{\frac{k_S(m_1 + m_2) \pm \sqrt{k_S^2(m_1 + m_2)^2 - 4m_1 m_2 k_S^2 \sin^2(\kappa\frac{d}{2})}}{m_1 m_2}}. \tag{3.44}$$

Defining $\omega_1 = \sqrt{k_S/m_1}$, $\omega_2 = \sqrt{k_S/m_2}$, and $\omega_0 = \sqrt{\omega_1^2 + \omega_2^2}$, we notice that there are two ω/ω_0 solutions for the ranges $[0, \ \omega_1/\omega_0]$ and $[\omega_2/\omega_0, \ \sqrt{2}]$ (see Fig. 3.13). The solution in the first range is usually known as the *acoustic* branch, while the second the *optical* branch. We notice the appearance of a band gap for frequencies between ω_1/ω_0 and ω_2/ω_0, and for all the frequencies above $\sqrt{2}$. As for the monoatomic case, the amount of attenuation can be obtained by solving the dispersion relation at a given frequency in terms of the wavenumber:

$$\kappa d = 2\arcsin\left(\sqrt{\frac{k_S^2(m_1 + m_2)^2 - [\omega^2 m_1 m_2 - k_S(m_1 + m_2)]^2}{4m_1 m_2 k_S^2}}\right). \tag{3.45}$$

It is important to notice that the addition of different masses increases the number of dispersion branches, due to the enriched dynamics of the system. We now look at the transmission function for the case of a single cell (Fig. 3.14a), and different number of cells (Fig. 3.14b). We see that the highest band gap opens always at $\beta = \sqrt{2}$ (Fig. 3.14a). In addition, increasing the number of cells, the level of attenuation increases, and other modes appear, but always below $\beta = \sqrt{2}$ (Fig. 3.14b).

We now move to an aperiodic beam composed of different materials, or a cross section variation that can be modeled as a random spring-mass chain (Fig. 3.15).

The characterization of aperiodic chains is more complex and they are usually investigated using the Anderson theory of localization [17, 18]. For the sake of simplicity, we analyse these structures by looking at the transmission functions only.

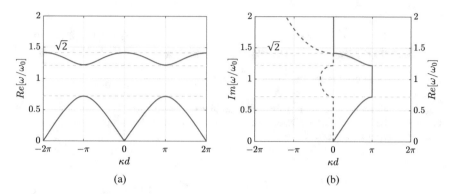

Fig. 3.13 **a** Dispersion relation for longitudinal waves in an inhomogeneous beam modeled through a spring-mass chain with equivalent elastic constants, given by an elastic spring and two point masses. **b** Real and imaginary part of the dispersion relation

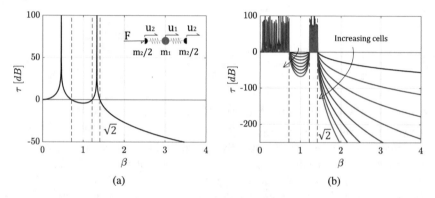

Fig. 3.14 **a** Transmission function for a single cell made of three masses connected with two springs (see inset) and **b** for different number of cells. Increasing the number of cells, the attenuation increases. A band gap exists in the interval $[\omega_1/\omega_0, \ \omega_2/\omega_0]$ and $[\sqrt{2}, \ +\infty)$

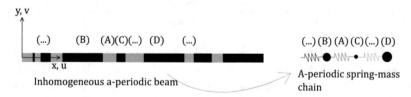

Fig. 3.15 Inhomogeneous beam with spatial aperiodic modulation of the elastic properties, and aperiodic spring-mass chain idealization

The proposed results are not an exhaustive interpretation of aperiodicity, but should be considered as examples to give an heuristic interpretation of the effects of randomisation in the spring-mass chain dynamics. Specifically, we describe aperiodic structures comparing them with periodic chains. We consider, as reference, a periodic chain composed of n masses (i.e. $(n-1)$ cells), and we start introducing small perturbations in both masses and springs, with the constraint of preserving the total mass and stiffness of the system. Specifically, if we define m_{iP} and k_{iP} the periodic ith mass and stiffness, and m_{iA} and k_{iAp} the aperiodic ones, we have:

$$m_{tot.} = \sum_{i=1}^{n} m_{iP} = (n-1)\bar{m} = \sum_{i=1}^{n} m_{iA},\qquad(3.46)$$

$$k_{tot.} = \left[\sum_{i=1}^{n-1} \frac{1}{k_{iP}}\right]^{-1} = \frac{\bar{k}}{n-1} = \left[\sum_{i=1}^{n-1} \frac{1}{k_{iA}}\right]^{-1},\qquad(3.47)$$

where $\bar{m}$ and $\bar{k}$ denote the mass and stiffness of each periodic cell (with the first and the last mass of the chain being equal to $\bar{m}/2$). We see that small perturbations in masses and springs (Fig. 3.16b) in a given chain (9 cells in the proposed example) with respect to the periodic case, does not affect significantly the attenuation capabilities of the system (Fig. 3.16a). A very comparable band gap opening position can be observed, as well as a comparable level of attenuation (Fig. 3.16a).

This example corroborates the claim that the band gap, in finite spring-mass chains, is not due to periodicity but discretisation. The discretization of the system gives rise to a discretised dynamic behaviour, with a subsequently limited number of vibration modes, able to support the elastic waves only for limited frequency intervals. After the last vibration mode, no propagating waves are supported. On the other hand, it is important to notice that the randomisation (even if small) introduces small attenuation bands in the transmission function before $\beta = \sqrt{2}$ (Fig. 3.16a).

In order to investigate the effect of randomisation on the attenuation capabilities of the chain, we keep for simplicity constant stiffness (as in the periodic chain) and change the masses only. Specifically, we introduce a mass distribution with increased variance (Fig. 3.17b) with respect to the previous case. Looking at the transmission function we can notice that (Fig. 3.17a):

• the aperiodic chain begins to attenuate before the value of $\beta = \sqrt{2}$;
• the aperiodic chain is endowed with transmission peaks after the value of $\beta = \sqrt{2}$, but inside a global attenuation trend.

This behaviour is a consequence of the combination of cells with different dynamics. Specifically, the constraint of having the same global mass/stiffness, implies that the local mass/stiffness values are forced both to increase and decrease with respect to the periodic case. This results in a distribution of the cells' resonances before and after the one of the periodic system.

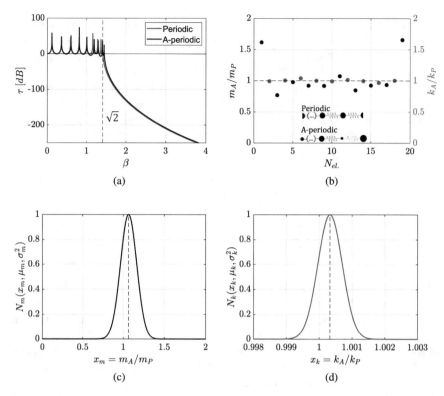

Fig. 3.16 **a** Transmission function for a periodic and aperiodic chain with equal global mass, stiffness and similar band gap opening position. **b** Mass and stiffness values in the aperiodic chain with respect to the periodic one, and corresponding gaussian distributions for the mass (**c**) and stiffness (**d**)

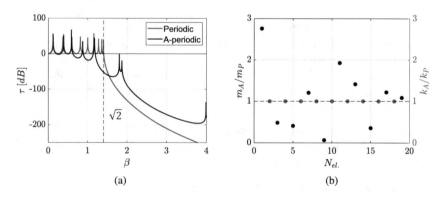

Fig. 3.17 **a** Transmission function for a periodic and aperiodic chain with equal global mass and stiffness. **b** Mass and stiffness values in the aperiodic chain with respect to the periodic one

If the resonance frequency of a cell is reduced, this imply that this cell will attenuate first; on the contrary, if it is increased, it will attenuate later. The anticipated attenuation trend is given by the cells with lower resonance frequency (higher mass or lower stiffness) while the transmission peaks after $\beta = \sqrt{2}$ by the cells with higher resonance frequency (lower mass or higher stiffness). On the other hand, if the variance is very small (i.e. the mass/stiffness fluctuations are very small) this effect is almost negligible (see Fig. 3.16a). It is now interesting to notice the effect of different distributions with respect to the random case. Specifically, we consider the simplest cases of step and linear mass distributions. It is clear that, with respect to the periodic chain, this system has a nonreciprocal behaviour, i.e. the response in one direction is different with respect to the one along the other. We see that the *graded* linear system is the one which provides an attenuation closest to the periodic chain (see Fig. 3.18c, g), with also an anticipation of the attenuation trend. This preliminary shows that the grading of inertial or elastic parameters is a powerful tool to manipulate elastic waves.

In addition, the chains with higher masses at the end (Fig. 3.18e–h) have a better attenuation performance with respect to the ones with lower masses at the end (Fig. 3.18a–d) (notice that the mass ratio of the first and last mass of the chain are just twice/half the others because of the halved initial and final periodic chain mass). In order to emphasise this nonreciprocal effect, we directly compare in Fig. 3.19a, b the step and linear mass distribution along the two different directions. For both cases, for attenuation purposes, it is preferable to put the higher masses at the end of the chain and not at the beginning, as also motivated by the previous considerations on phases and Argand diagrams. In this way it is possible to anticipate the opening of the attenuation range, as well as to increase the global level of attenuation (Fig. 3.19a, b).

Experiments on aperiodic structures. We now conclude the investigation of aperiodic structures proposing an experimental validation of their attenuation capabilities. Specifically, we compare a periodic and aperiodic structure that can be described using simple spring-mass chain models as in [3]. Both systems are analysed numerically (FEM) and experimentally through a comparison of the corresponding transmission functions. The aperiodic structure has been defined by perturbating an already existing periodic one [15]. Prototypes are built using Selective Laser Sintering (SLS) Additive Manufacturing (AM) technology. Both periodic and aperiodic structures define a slice with 15×15 cm^2 in-plane dimensions. Externally, the two structures are identical, with semispheres of the same dimensions, while internally, the aperiodic one (Fig. 3.20b) is significantly perturbed with respect to the periodic one (Fig. 3.20a). The internal random perturbation preserves the same unit cell concept with four spheres attached on a frame. Details on the geometry are reported in Table 3.1 with reference to the unit cell side length $a = 5$ cm. The adopted material is Nylon PA2200 (Young's modulus $E = 1.70$ GPa, Poisson's ratio $\nu = 0.4$ and density $\rho = 925$ kg/m^3.

The two structures have different total mass and stiffness, as shown in Table 3.2.

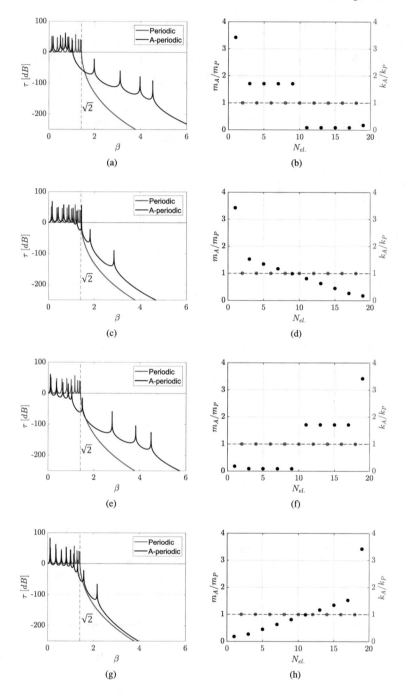

Fig. 3.18 Transmission function for decreasing (**a**, **c**) and increasing (**e**, **g**) step and linear mass functions respectively and corresponding distributions (**b**, **d**) and (**f**, **h**)

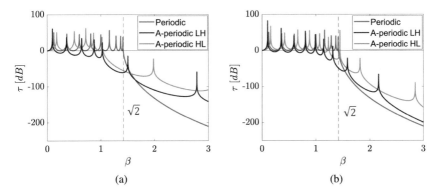

Fig. 3.19 **a** Transmission function for periodic and aperiodic cases with step mass distribution from lower to higher (LH) and higher to lower (HL) masses. Resonance peaks and attenuation regions are alternated for LH and HL. **b** Same concept using a linear mass distribution

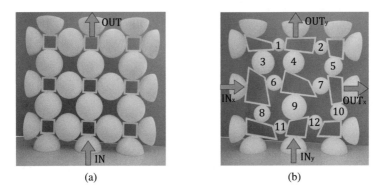

Fig. 3.20 Analysed **a** periodic and **b** aperiodic structures obtained through SLS additive manufacturing technology

Table 3.1 Geometric parameters. $w_{frame}^{in-pl.}$ and $w_{frame}^{out-pl.}$ are the in-plane and out-of-plane frame thicknesses

	PERIODIC (Fig. 1a)	APERIODIC (Fig. 1b)
$w_{frame}^{in-pl.}$	0.04a	0.04a
$w_{frame}^{out-pl.}$	0.05a	0.05a
r_{sph}	0.33a	(1) 0.152a, (2) 0.167a, (3) 0.279a, (4) 0.265a, (5) 0.223a, (6) 0.197a, (7) 0.210a, (8) 0.207a, (9) 0.311a, (10) 0.208a, (11) 0.173a, (12) 0.180a

Table 3.2 Global values of stiffness and mass computed numerically (FEM)

Structure	$k^x_{glob.}$ (kN/m)	$k^y_{glob.}$ (kN/m)	$m_{glob.}$ (g)
PERIODIC	95.5	95.5	316
APERIODIC	30.2	28.8	174
$\Delta\%_{100[1-(a-per./per.)]}$	68.4	69.8	44.9

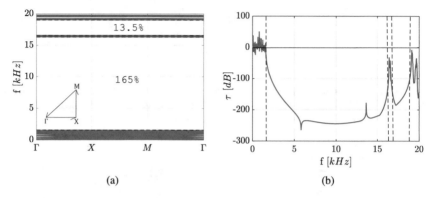

(a)   (b)

Fig. 3.21 **a** Numerical (FEM) dispersion and **b** transmission for the periodic structure. Band gap width computed using the gap-midgap ratio

The dynamic behavior of the periodic structure is described via a linear elastic dispersion analysis. The numerical band structure is calculated by means of Abaqus and a proper implemented routine to apply Bloch–Floquet boundary conditions. The dispersion relation and transmission function of the periodic structure are reported in Fig. 3.21, where a very wide band gap in the frequency range of 1.57–16.29 kHz can be observed.

The periodic and aperiodic structures are compared looking at their linear elastic numerical transmission functions between the input and output points shown in Fig. 3.20. Figure 3.22a shows the transmission functions in the range of 0–20 kHz: both structures exhibit high attenuation properties in almost the same frequency range even though the aperiodic structure involves a lot of moderate peaks due to the presence of many local modes.

It is worth noticing that attenuation gaps appear in the aperiodic structure at low frequencies as shown in Fig. 3.22b. The associated deformed configurations involve local movements of internal portions of the structure of both spheres and frames. Specifically, the aperiodic structure exhibits a stronger attenuation at low frequencies: this is attributed to the local lower stiffness, as well as higher masses along the transmission line. This behaviour totally comply with previous spring-mass chain predictions since we see both the anticipation of the attenuation range, and transmission peaks inside the periodic structure band gap.

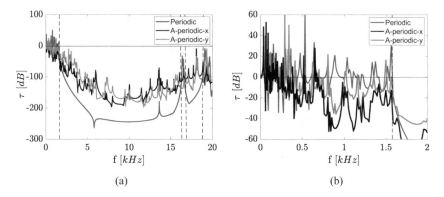

Fig. 3.22 **a** Numerical (FEM) transmission functions in the frequency range of 0–20 kHz and **b** detailed view in the frequency range of 0–2 kHz

Experimental tests are carried out in order to validate the numerical results. Prototypes are built using SLS Additive Manufacturing technology. The experimental setup is composed of an inertial shaker LDSv406 connected to a PA 100E Power Amplifier and two PCB Piezotronics 353B15 accelerometers, with a sensitivity of 10 mV/g and a resonance frequency of 70 kHz. A white noise excitation is imposed to the shaker. The input acceleration is measured using an accelerometer glued between the prototype and the shaker connection, while the output acceleration is measured at the opposite face of the periodic/aperiodic structure. The structure is placed on a very soft rubber foam, to avoid disturbances between input and output (see Fig. 3.23). Data acquisition is performed using NiMax Measurement Automatic explorer and postprocessed by means of a user Matlab routine. In Figs. 3.24, 3.25, 3.26, numerical and experimental results are compared in the frequency range of 0–20 kHz and with a detailed view in the range of 0–3 kHz. Experimental results are in good agreement with numerical simulations, confirming the high attenuation capabilities of both periodic and aperiodic structures. Differences in attenuation are related to the accelerometer sensitivity, which provides a cut-off level of attenuation at around −60 dB.

This is also demonstrated by looking at the coherence functions (Fig. 3.27) that are equal to zero in the band gap regions, and are defined as:

$$Coh_{XY}(f) = \frac{S_{XY}(f)}{\sqrt{S_X(f)S_Y(f)}}, \qquad (3.48)$$

where $S_X(f)$ and $S_Y(f)$ are the power spectral densities (S.P.D.) and and $S_{XY}(f)$ the cross power spectral density (C.S.P.D.), respectively equal to the Fourier transform of the autocorrelation and cross-correlation functions (Wiener–Khintchine theorem). The examination of the results in the low-frequency regime confirms that the periodic structure shows an almost continuous transmission until the abrupt reduction in correspondence with the band gap opening. On the other hand, the aperiodic structure

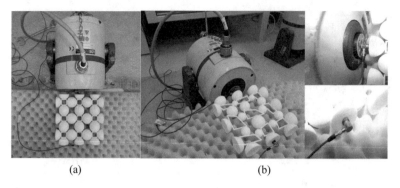

(a)   (b)

Fig. 3.23 **a** Experimental setup for the periodic (**a**) and aperiodic (**b**) structure, and corresponding details of the accelerometers glued at the center of the top and bottom faces

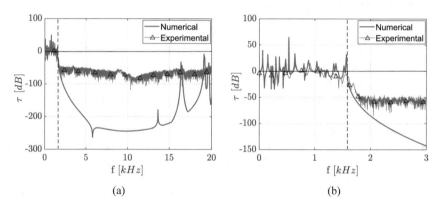

(a)   (b)

Fig. 3.24 **a** Numerical (FEM) and experimental transmission function of the periodic structure in the frequency range of 0–20 kHz and **b** detailed view in the frequency range 0–3 kHz

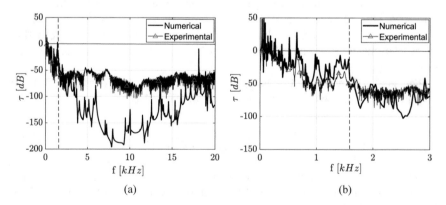

(a)   (b)

Fig. 3.25 **a** Numerical (FEM) and experimental transmission function for the aperiodic structure along x in the frequency range of 0–20 kHz and **b** detailed view in the frequency range 0–3 kHz

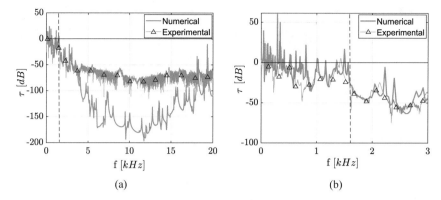

Fig. 3.26 **a** Numerical (FEM) and experimental transmission function for the aperiodic structure along y in the frequency range of 0–20 kHz and **b** detailed view in the frequency range 0–3 kHz

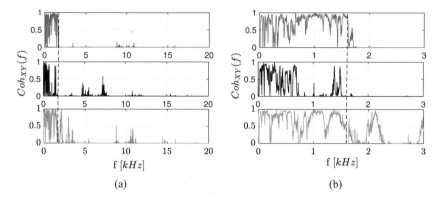

Fig. 3.27 **a** Experimental coherence function in the frequency range of 0–20 kHz and **b** detailed view in the frequency range 0–3 kHz, for the periodic (red) and aperiodic structure in x (black) and y direction (grey) respectively

starts to attenuate earlier, but several peaks are experimentally denoted, especially for the y direction. This is in agreement with the numerical results and with the theoretical predictions for random spring-mass chains.

3.4 Physical Interpretation of Local Resonance

We can simplify the local resonance phenomenon by looking again at the spring-mass chains proposed in the previous section for the behaviour of inhomogeneous media. Specifically, the phenomenon can be explained by changing the perspective of observation: while before we observed the dynamics of the structure between the

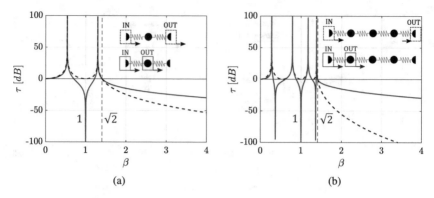

Fig. 3.28 Transmission functions for a spring mass chain composed of two (**a**) and four (**b**) cells. By measuring the output on the second mass of the first cell, we notice a typical locally resonant behaviour

initial and final cell of the chain, now we focus on the first cell only. The concept is schematically explained in Fig. 3.28.

By looking at the displacement field of the second mass of the first unit cell (Fig. 3.28a), it can be noticed that an anti-resonance appears for $\beta = 1$, meaning that the system enters in attenuation regime before the well known value of $\beta = \sqrt{2}$. The anti-resonance is associated to the resonance of the last mass at $\omega_0 = \sqrt{2k/m}$, and appears in order to guarantee the energy balance. A similar concept can be noticed when more cells are added (see Fig. 3.28b), which is equivalent to consider more vibration modes of the resonating element. We now simplify the system, considering only two masses and two springs (see inset in Fig. 3.29), in order to investigate the effect of different mass and stiffness ratios. The equations of motions are defined as:

$$\begin{cases} m_0\ddot{u}_1(t) + k_0[u_1(t) - u_0(t)] + k_R[u_1(t) - u_R(t)] = 0, \\ m_R\ddot{u}_2(t) + k_R[u_2(t) - u_1(t)] = 0. \end{cases} \tag{3.49}$$

where 0 denotes the main structure with displacement u_1 and R the resonator with displacement u_2. Substituting again the plane wave solution of Eq. (3.5), with some algebraic manipulations the following is obtained:

$$\begin{cases} u_1(t) = \dfrac{k_0(k_R - \omega^2 m_R)}{(-m_0\omega^2 + k_0 + k_R)(k_R - \omega^2 m_R) - k_R^2} u_0 e^{i\omega t}, \\ u_2(t) = \dfrac{k_0 k_R}{(-m_0\omega^2 + k_0 + k_R)(k_R - \omega^2 m_R) - k_R^2} u_0 e^{i\omega t}. \end{cases} \tag{3.50}$$

By defining $\beta = \omega/\omega_R$, with $\omega_R = \sqrt{k_R/m_R}$, it can be noticed that increasing the ratio m_R/m_0, the upper and lower band gap limits shift to higher β (Fig. 3.29a), globally enlarging the band gap width until a convergence value. On the other hand,

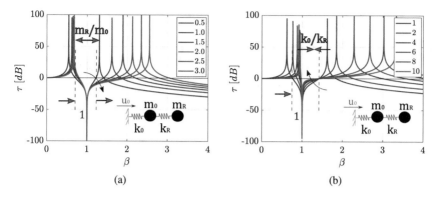

Fig. 3.29 Transmission function in a 2 dof locally resonant structure for different mass (**a**) and stiffness (**b**) ratios between the resonator and the main structure

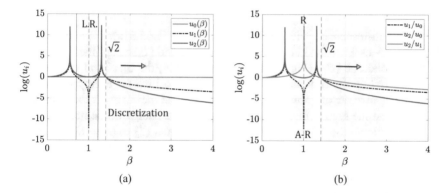

Fig. 3.30   a Displacement field of the input u_0, first mass u_1 and second mass u_2 for a 2 dof locally resonant structure, and **b** corresponding displacement ratios, i.e. transmission functions

by increasing the ratio k_0/k_R, the lower band gap limit shift to higher β, while the upper one remains fixed, with a global reduction of the attenuation levels, reaching in the limit case a sharp resonance and anti-resonance at $\beta = 1$ (Fig. 3.29b).

In order to understand the local resonance (L.R.) physics, we investigate the displacement field of the two masses (with $m_R = m_0/2$) and draw some energy and phase considerations by means of Argand diagrams. It is worth to notice that, for a constant input displacement u_0, the second mass exhibits only a band gap due to discretization (Fig. 3.30a), while the first one two, due to L.R. and discretization. By looking at the displacement ratios, it can be noticed that the L.R. band gap is related to the resonance of the second mass (Fig. 3.30b). This suggests that the attenuation in the first mass is due to an amplification in the second mass, i.e. the energy remains localised in the resonator.

In order to justify this energy statement, we directly look at the energy quantities inside the L.R. band gap, for different values of β, as shown in Fig. 3.31. Specifically we investigate the energy distribution in significant points related to the band gap

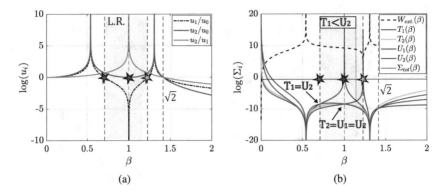

Fig. 3.31 Zoomed view of the displacement ratios in the band gap region (**a**) and corresponding energy distributions in a 2 dof locally resonant structure

opening, anti-resonance, and band gap closing (see Fig. 3.31a). We notice that the band gap opens when the kinetic energy of the first cell (T_1) equals the strain energy of the second one (U_2) (Fig. 3.31b). After this point, the first band gap region is characterized by the condition $T_1 < U_2$ (we take positive the energy given to the system, negative the one absorbed), which means that the energy is localised in the resonator. At the second mass resonance, the kinetic energy of the first mass goes to zero, while all the other energy quantities are equal ($T_2 = U_1 = U_2$). The band gap closure is obtained when the strain energy of the first cell U_1 goes to zero. As expected, for $\beta > \sqrt{2}$, we get the conventional band gap due to the discretization (i.e. inhomogeneity) of the system.

It is clear that the most important parameters are the main structure kinetic energy T_1, and the resonator internal energy U_2. We then define a localised energy as:

$$\Sigma_{loc.} = T_1 + U_2 \tag{3.51}$$

which represents the energy of the system except from the input and output regions (strain energy of the first spring k_0 and kinetic energy of the last mass m_R). The difference between this quantity and the external work ($\Delta\Sigma = \Sigma_{loc.} - W_{ext.}$) shows that the band gap opening and closure positions are related to its minimum and maximum, while for $\beta = 1$ all the energy is localized ($\Sigma_{loc} = W_{ext}$) (Fig. 3.32a). In addition, we notice that the minimum value in the closure band gap position is due to the lower external work (which tends to zero), which suggests an out-of-phase response. A dual perspective is based on the observation of the non-localised energy $T_2 + U_1$, with a maximum in the band gap opening position and minimum in the closure band gap position (Fig. 3.32b).

We now complement the energy interpretation by looking at the phases through Argand diagrams. We notice that the band gap opens with an in-phase response (Fig. 3.33a, b). At the anti-resonance the displacement of the first mass goes to zero, with an out-of-phase response Fig. 3.33b, c. Band gap closure is like the opening, but

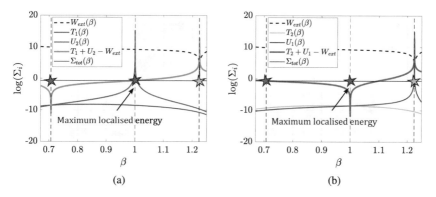

Fig. 3.32 Localised (**a**) and non-localised (**b**) energy distributions inside a 2 dof locally resonant band gap

with an out-of-phase response Fig. 3.33e, f. We now consider the limit cases of high mass and stiffness ratios ($m_R/m_0 = k_0/k_R = 50$), looking both at the displacement and energy diagrams. In the case of high mass ratios, the two band gaps merge, with an opening position at $\beta = 1$ (Fig. 3.34a). In terms of energy, the second peak of the internal energy U_2, related to the band gap closure, disappears (Fig. 3.34b). Contrary, by increasing the stiffness of the main structure with respect to the resonator, the L.R. band gap almost disappears, having attenuation only after $\beta = \sqrt{2}$ (Fig. 3.35a). In terms of energy, the kinetic energy peak of the first mass T_1 becomes really sharp with almost coincident resonance and anti-resonance, and the internal energy peak moves to $\beta = \sqrt{2}$ (Fig. 3.35b) like in a conventional inhomogeneous medium.

We now move to periodic structures made of locally resonant unit cells (see inset in Fig. 3.36a). We consider for simplicity a cell made of a lumped spring-mass, with attached a lumped resonator. The equations of motion are:

$$\begin{cases} m_0 \ddot{u}_n(t) + k_0[u_n(t) - u_{n-1}(t)] + k_0[u_n(t) - u_{n+1}(t)] + k_R[u_n(t) - u_{Rn}(t)] = 0, \\ m_R \ddot{u}_{Rn}(t) + k_R[u_{Rn}(t) - u_n(t)] = 0. \end{cases}$$

$$(3.52)$$

Substituting the plane wave solution defined in Eq. (3.5) and imposing periodicity conditions, we obtain:

$$\begin{cases} [-\omega^2 m_0 + 2k_0]\tilde{u}_n - k_0[e^{-i\kappa d} + e^{i\kappa d}]\tilde{u}_n + k_R[\tilde{u}_n - \tilde{u}_{Rn}] = 0, \\ [-\omega^2 m_R + k_R]u_{Rn} - k_R u_n = 0. \end{cases}$$

$$(3.53)$$

From which, non-trivial solutions are possible only for vanishing determinant, i.e.

$$\cos(\kappa d) = 1 - \frac{m_0}{2k_0}\omega^2 + \frac{k_R}{2k_0}\left(1 - \frac{k_R}{k_R - \omega^2 m_R}\right). \tag{3.54}$$

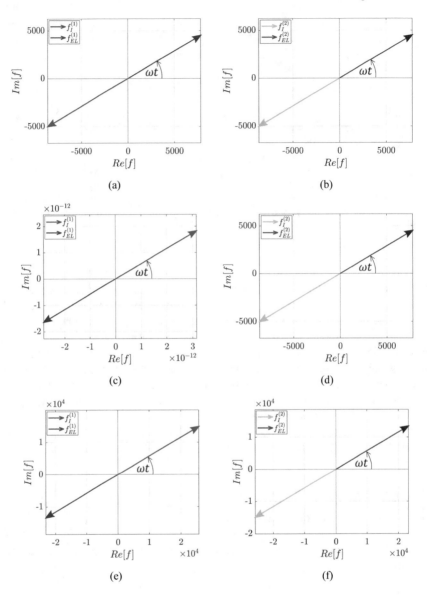

Fig. 3.33 Argand diagrams for different values of β. First and second mass for $\beta = 0.7071$ (L.R. band gap opening) (**a**, **b**), anti-resonance $\beta = 1$ and $\beta = 1.224$ (L.R. band gap closure) (**e**, **f**)

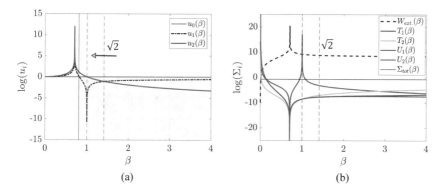

Fig. 3.34 **a** Displacement field of the input u_0, first mass u_1, and second mass u_2 for a 2 dof locally resonant structure with $m_R/m_0 = 50$, and **b** corresponding energy distributions

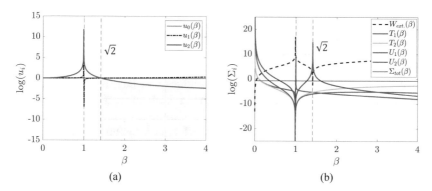

Fig. 3.35 **a** Displacement field of the input u_0, first mass u_1, and second mass u_2 for a locally resonant structure with $k_0/k_R = 50$, and **b** corresponding energy distributions

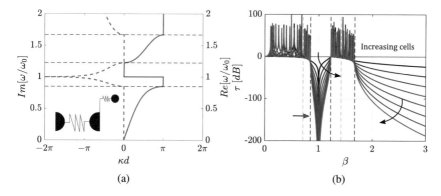

Fig. 3.36 **a** Dispersion relation for a spring-mass chain with resonators ($m_R/m_0 = 0.5$ and $k_R = k_0$) and **b** corresponding transmission functions for different number of cells. Increasing the number of cells, the attenuation increases, and the L.R. band gap moves to higher frequencies

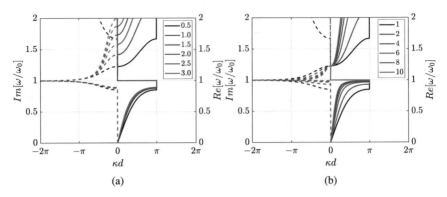

Fig. 3.37 Dispersion relation for a spring-mass chain with resonators for different m_R/m_0 (**a**) and k_0/k_R (**b**) ratios

The first contribution in Eq. (3.54) defines the dispersive behaviour of the spring-mass chain support (as in Eq. (3.5)), while the second one is due to the resonators. In Fig. 3.36, it is reported the dispersion relation and the transmission functions for different number of cells. It is important to notice that increasing the number of cells, the band gap moves to higher β (like in Fig. 3.29) due to a modification of the global mass and stiffness. On the other hand, as expected, by increasing the number of cells, the attenuation level increases. The effect of different m_R/m_0 and k_0/k_R ratios for periodic systems, can be analysed by looking at the real and imaginary part of the dispersion curves (Fig. 3.37). It can be noticed that the band gap increases for increasing resonating mass (Fig. 3.37a), while increasing the stiffness of the main structure it remains constant (Fig. 3.37b) but the imaginary part shrinks (i.e. the attenuation decreases). By defining $\mu = \kappa d$, $\Omega^2 = k_0/m_0$, $\omega_R^2 = k_R/m_R$ and the ratios $\bar{\omega}^2 = \omega^2/\Omega^2$, $\bar{\omega}^2 = \omega^2/\Omega^2$, $\bar{\omega}_R^2 = \omega_R^2/\Omega^2$ we get:

$$\cos(\mu) = 1 - \frac{\bar{\omega}^2}{2}\left[1 + \frac{m_R}{m_0}\frac{\bar{\omega}_R^2}{\bar{\omega}_R^2 - \bar{\omega}^2}\right]. \tag{3.55}$$

Evanescent waves exist if (i) $\cos(\mu) > 1$ and (ii) $\cos(\mu) < -1$.
Condition (i) gives:
if $\bar{\omega}^2 < \bar{\omega}_R^2 \rightarrow \bar{\omega}^2 > \bar{\omega}_R^2\left(1 + \frac{m_R}{m_0}\right) \rightarrow$ impossible;
if $\bar{\omega}^2 > \bar{\omega}_R^2 \rightarrow \bar{\omega}^2 < \bar{\omega}_R^2\left(1 + \frac{m_R}{m_0}\right)$.
Condition (ii) gives:
if $\bar{\omega}^2 < \bar{\omega}_R^2 \rightarrow \bar{\omega}^4 + \bar{\omega}^2(-\bar{\omega}_R^2\frac{m_R}{m_0} - 4 - \bar{\omega}_R^2) + 4\bar{\omega}_R^2 < 0$, whose solutions are:

$$\bar{\omega}_1^2 = \frac{4 + \bar{\omega}_R^2\left(1 + \frac{m_R}{m_0}\right) - \sqrt{\left[4 + \bar{\omega}_R^2\left(1 + \frac{m_R}{m_0}\right)\right]^2 - 16\bar{\omega}_R^2}}{2} < \bar{\omega}^2 < \frac{4 + \bar{\omega}_R^2\left(1 + \frac{m_R}{m_0}\right) + \sqrt{\left[4 + \bar{\omega}_R^2\left(1 + \frac{m_R}{m_0}\right)\right]^2 - 16\bar{\omega}_R^2}}{2} =$$

$\bar{\omega}_2^2$. In order to comply with the hypothesis $\bar{\omega}^2 < \bar{\omega}_R^2$, we need to verify that $(\bar{\omega}_1^2, \bar{\omega}_2^2) < \bar{\omega}_R^2$, obtaining that $\bar{\omega}_1^2 < \bar{\omega}^2 < \bar{\omega}_R^2$.
If $\bar{\omega}^2 > \bar{\omega}_R^2 \rightarrow \bar{\omega}^2 > \bar{\omega}_2^2 > \bar{\omega}_R^2$.

In conclusion, evanescent waves exist in the range:
$\left(\bar{\omega}_1^2; \ \bar{\omega}_R^2(1 + \frac{m_R}{m_0})\right)$ and $\left(\bar{\omega}_2^2; \ +\infty\right)$.
This implies that the L.R. band gap is defined as:

$$\frac{4\Omega^2 + \omega_R^2\left(1 + \frac{m_R}{m_0}\right) - \sqrt{\left[4\Omega^2 + \omega_R^2\left(1 + \frac{m_R}{m_0}\right)\right]^2 - 16\omega_R^2\Omega^2}}{2} < \omega^2 < \omega_R^2\left(1 + \frac{m_R}{m_0}\right).$$

$$(3.56)$$

In conclusion, we notice that one-dimesional models for periodic and aperiodic structures are able to provide essential insights into the most common linear wave propagation mechanisms. Specifically, the concept of dispersion curve, band gap and local resonance will be extensively used in the next chapters. Moreover, the energy and phase arguments show that local resonance is a good candidate to simultaneously provide vibration insulation and energy localisation in the resonators.

References

1. J.D. Achenbach, S.A. Thau, Wave propagation in elastic solids. J. Appl. Mech. (1974)
2. L. Brillouin, Wave propagation in periodic structures: electric filters and crystal lattices. Nature (1946)
3. J.M. De Ponti, N. Paderno, R. Ardito, F. Braghin, A. Corigliano, Experimental and numerical evidence of comparable levels of attenuation in periodic and a-periodic metastructures. Appl. Phys. Lett. (2019)
4. B.K. Henderson, K.I. Maslov, V.K. Kinra, Experimental investigation of acoustic band structures in tetragonal periodic particulate composite structures. J. Mech. Phys. Solids (2001)
5. O.R. Bilal, M.I. Hussein, Ultrawide phononic band gap for combined in-plane and out-of-plane waves. Phys. Rev. E Stat. Nonlinear Soft Matter Phys. (2011)
6. S. Babaee, P. Wang, K. Bertoldi, Three-dimensional adaptive soft phononic crystals. J. Appl. Phys. (2015)
7. V. Laude, *Phononic Crystals*, De Gruyter edn. (DE GRUYTER, Berlin, 2015)
8. L. D'Alessandro, E. Belloni, R. Ardito, A. Corigliano, F. Braghin, Modeling and experimental verification of an ultra-wide bandgap in 3D phononic crystal. Appl. Phys. Lett. **109**(22), 1–5 (2016)
9. L. D'Alessandro, E. Belloni, R. Ardito, F. Braghin, A. Corigliano, Mechanical low-frequency filter via modes separation in 3D periodic structures. Appl. Phys. Lett. **111**(23) (2017)
10. L. D'Alessandro, V. Zega, R. Ardito, A. Corigliano, 3D auxetic single material periodic structure with ultra-wide tunable bandgap. Sci. Rep. **8**(1), 1–9 (2018)
11. L. D'Alessandro, R. Ardito, F. Braghin, A. Corigliano, Low frequency 3D ultra-wide vibration attenuation via elastic metamaterial. Sci. Rep. (2019)
12. J.M. De Ponti, E. Riva, R. Ardito, F. Braghin, A. Corigliano, Wide low frequency bandgap in imperfect 3D modular structures based on modes separation. Mech. Res. Commun. (2020)
13. G.A. Gazonas, D.S. Weile, R. Wildman, A. Mohan, Genetic algorithm optimization of phononic bandgap structures. Int. J. Solids Struct. (2006)
14. H.W. Dong, X.X. Su, Y.S. Wang, C. Zhang, Topology optimization of two-dimensional asymmetrical phononic crystals. Phys. Lett. Sect. A: Gen. At. Solid State Phys. (2014)
15. L. D'Alessandro, B. Bahr, L. Daniel, D. Weinstein, R. Ardito, Shape optimization of solid–air porous phononic crystal slabs with widest full 3D bandgap for in-plane acoustic waves. J. Comput. Phys. **344**, 465–484 (2017)

16. J. Vondřejc, E. Rohan, J. Heczko, Shape optimization of phononic band gap structures using the homogenization approach. Int. J. Solids Struct. (2017)
17. S. Russ, Scaling of the localization length in linear electronic and vibrational systems with long-range correlated disorder. Phys. Rev. B Condens. Matter Mater. Phys. (2002)
18. F.A.B.F. De Moura, L.P. Viana, A.C. Frery, Vibrational modes in aperiodic one-dimensional harmonic chains. Phys. Rev. B Condens. Matter Mater. Phys. (2006)

Chapter 4
Graded Elastic Metamaterials

Abstract This chapter analyses the problem of wave propagation in elastic continua, with specific reference to plates and half-spaces. Wave propagation in thin elastic plates and half-spaces is studied going from classic analytical theories to numerical (FEM) models able to describe the interaction of waves in such systems with arrays of resonators. The concept of grading is then introduced as a way to obtain broadband band gaps and the rainbow effect, i.e. a spatial signal separation depending on frequency. Finally, reversed mode conversion from surface Rayleigh to Shear (S) and Pressure (P) bulk waves is demonstrated leveraging on the Umklapp phenomenon. This mechanism allows to manipulate surface waves, focusing the elastic energy in specific regions for a broadband input frequency.

4.1 Problem Statement in Dynamic Elasticity

In the setting of elasticity, we start investigating the problem of elastic wave propagation in plates and half-spaces, with specific reference to the mathematical treatise in [1], generalizing the previously obtained formulas obtained for 1D elastic wave propagation in homogeneous media.

Kinematics. Let us define the displacement field of particles as $\mathbf{u}(\mathbf{x}, t)$. Within the restriction of linearized theory, the deformation is described using the small-strain tensor:

$$\varepsilon_{ij} = \frac{1}{2}(u_{i,j} + u_{j,i}), \tag{4.1}$$

which is a symmetric ($\varepsilon_{ij} = \varepsilon_{ji}$) rank two tensor. We define also the rotation tensor as:

$$\omega_{ij} = \frac{1}{2}(u_{i,j} - u_{j,i}), \tag{4.2}$$

which is antisymmetric ($\omega_{ij} \neq \omega_{ji}$).

Equilibrium. Suppose now we remove from a body of volume Ω and mass density ρ a closed region $S + V$, where S is the boundary as in Fig. 4.1.

© The Author(s), under exclusive license to Springer Nature Switzerland AG 2021
J. M. De Ponti, *Graded Elastic Metamaterials for Energy Harvesting*,
PoliMI SpringerBriefs, https://doi.org/10.1007/978-3-030-69060-1_4

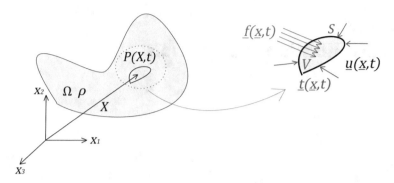

Fig. 4.1 Schematic of the elastic continuum used to write the elastodynamic equilibrium equations

The surface S is subjected to a distribution of surface tractions $\mathbf{t}(\mathbf{x}, t)$ and each mass element to a body force per unit mass $\mathbf{f}(\mathbf{x}, t)$. The principle of conservation of linear momentum in the linearized theory reads:

$$\int_S \mathbf{t} dA + \int_V \rho \mathbf{f} dV = \int_V \rho \ddot{\mathbf{u}} dV. \tag{4.3}$$

Using the Cauchy stress formula, $t_l = \sigma_{kl} n_k$, we get in index notation:

$$\int_S \sigma_{kl} n_k dA + \int_V \rho f_l dV = \int_V \rho \ddot{u}_l dV. \tag{4.4}$$

By using the Gauss' theorem, the surface integral can be transformed into a volume integral, obtaining:

$$\int_V (\sigma_{kl,k} + \rho f_l - \rho \ddot{u}_l) dV = 0. \tag{4.5}$$

Since V may be an arbitrary part of the body, it follows that:

$$\sigma_{kl,k} + \rho f_l - \rho \ddot{u}_l = 0, \tag{4.6}$$

which is the well known Cauchy's first law of motion.

Writing now the principle of conservation of the angular momentum we get:

$$\int_S (\mathbf{x} \wedge \mathbf{t}) dA + \int_V (\mathbf{x} \wedge \mathbf{f}) \rho dV = \int_V \rho \frac{\partial}{\partial t} (\mathbf{x} \wedge \dot{\mathbf{u}}) dV. \tag{4.7}$$

Simplifying the right-hand side and introducing index notation:

$$\int_S e_{klm} x_l t_m dA + \int_V e_{klm} x_l f_m dV = \int_V \rho e_{klm} x_l \ddot{u}_m dV. \tag{4.8}$$

Using again the Cauchy stress formula and Gauss' theorem:

$$\int_S e_{klm} x_l \sigma_{km} n_k dA = \int_V e_{klm} (\delta_{lk} \sigma_{km} + x_l \sigma_{km,k}) dV. \tag{4.9}$$

Using the first law of motion stated in Eq. (4.6) we get:

$$\int_V e_{klm} \delta_{lk} \sigma_{km} = 0, \tag{4.10}$$

which implies that:

$$\sigma_{lm} = \sigma_{ml} \tag{4.11}$$

i.e. the stress tensor is symmetric.

Constitutive law. It is possible to define the linear relation between stress and strain tensors as:

$$\sigma_{ij} = C_{ijkl} \varepsilon_{kl}. \tag{4.12}$$

where

$$C_{ijkl} = C_{jikl} = C_{klij} = C_{ijlk}. \tag{4.13}$$

Thus, 21 of the 81 components of the tensor are independent. If the coefficients C_{ijkl} are constants, the medium is homogeneous. On the other hand, the material is *elastically isotropic* when there are no preferred directions in the material, and the elastic constants are the same independently on the orientation of the coordinate system in which σ_{ij} and ε_{kl} are evaluated. For the case of elastic isotropy, the constants C_{ijkl} can be expressed as:

$$C_{ijkl} = \lambda \delta_{ij} \delta_{kl} + \mu (\delta_{ik} \delta_{jl} + \delta_{il} \delta_{jk}), \tag{4.14}$$

and Hooke's law assumes the form:

$$\sigma_{ij} = \lambda \varepsilon_{kk} \delta_{ij} + 2\mu \varepsilon_{ij} \tag{4.15}$$

where λ and μ are the well known Lamé coefficients, where $\lambda = Ev/[(1+v)(1-2v)]$ and $\mu = E/2(1+v)$.

Governing equations in dynamic elasticity. We consider a continuum with volume V (bounded or unbounded) with interior V, closure $\bar{V}$ and boundary S. The system of equations governing the motion of a homogeneous, isotropic, linear elastic body are:

$$\begin{cases} \sigma_{ij,j} + \rho f_i = \rho \ddot{u}_i \\ \sigma_{ij} = \lambda \varepsilon_{kk} \delta_{ij} + 2\mu \varepsilon_{ij} \\ \varepsilon_{ij} = \frac{1}{2}(u_{i,j} + u_{j,i}) \end{cases} \tag{4.16}$$

If the strain-displacement relations are substituted into Hooke's law and the expressions for the stresses are subsequently substituted in the stress-equations of motion, we obtain the displacement equations of motions:

$$\mu u_{i,jj} + (\lambda + \mu)u_{j,ji} + \rho f_i = \rho \ddot{u}_i \tag{4.17}$$

This system of equations has a disadvantageous feature in that it couples the three displacement components. A convenient approach is to express the components of the displacement vector in terms of derivatives of potentials. In vector notation, the displacement equation of motion can be written as:

$$\mu \nabla^2 \mathbf{u} + (\lambda + \mu)\nabla \nabla \cdot \mathbf{u} = \rho \ddot{\mathbf{u}}, \tag{4.18}$$

where $\nabla^2[\,]$ is the Laplacian:

$$\nabla^2[\,] = \frac{\partial^2[\,]}{\partial x_1^2} + \frac{\partial^2[\,]}{\partial x_2^2} + \frac{\partial^2[\,]}{\partial x_2^2}, \tag{4.19}$$

$\nabla[\,]$ the Gradient:

$$\nabla[\,] = i_1 \frac{\partial[\,]}{\partial x_1} + i_2 \frac{\partial[\,]}{\partial x_2} + i_3 \frac{\partial[\,]}{\partial x_3}, \tag{4.20}$$

and $\nabla \cdot [\,]$ the Divergence:

$$div[\,] = \nabla \cdot [\,] = \frac{\partial[\,]}{\partial x_1} + \frac{\partial[\,]}{\partial x_2} + \frac{\partial[\,]}{\partial x_2}. \tag{4.21}$$

We consider a decomposition of the displacement vector of the form:

$$\mathbf{u} = \nabla \varphi + \nabla \wedge \boldsymbol{\psi}. \tag{4.22}$$

Substituting Eq. (4.22) into Eq. (4.18) we get:

$$\nabla[(\lambda + 2\mu)\nabla^2 \varphi - \rho \ddot{\varphi}] + \nabla \wedge [\mu \nabla^2 \boldsymbol{\psi} - \rho \ddot{\boldsymbol{\psi}}] = 0. \tag{4.23}$$

The displacement representation of Eq. (4.22) satisfies the equation of motion if:

$$\nabla^2 \varphi = \frac{1}{c_L^2} \ddot{\varphi}, \tag{4.24}$$

$$\nabla^2 \boldsymbol{\psi} = \frac{1}{c_T^2} \ddot{\boldsymbol{\psi}}, \tag{4.25}$$

where:

$$c_L^2 = \frac{\lambda + 2\mu}{\rho} = \frac{E(1-\nu)}{\rho(1+\nu)(1-2\nu)}, \tag{4.26}$$

$$c_T^2 = \frac{\mu}{\rho} = \frac{E}{2\rho(1+\nu)}, \tag{4.27}$$

with c_L and c_T the longitudinal and transverse wave velocity respectively. Using the notation proposed at the beginning of the chapter, $\gamma_L(\nu) = \sqrt{(1-\nu)/[(1+\nu)(1-\nu)]}$, while $\gamma_T(\nu) = \sqrt{1/[2(1+\nu)]}$.

4.2 Wave Propagation in Thin Elastic Plates

The classical theory of bending of plates, is based on the *Love–Kirchhoff hypothesis*, originally formulated by Kirchhoff (1850) and adopted by Love (1927) in the *Mathematical Theory of Elasticity* [2]. They make reference to the straight segments normal to the midplane (generally called as normals) and state that in the deformed configuration they continue to be: (i) rectilinear, (ii) perpendicular to the middle surface and (iii) of unaltered length. These assumptions, which can be regarded as an extension to plates of the *plane section* hypothesis of Navier relative to beams, lead to a simple kinematical model for determining the displacement of a generic point across the thickness as a function of the deformed configuration of the middle surface. In addition, a direct consequence of the hypothesis (iii) is that stresses in the direction normal to the middle plane are neglected. With reference to Fig. 4.2, the displacement field is defined as:

$$\begin{cases} u(x, y, z, t) = z \cdot \varphi_{xz}(x, y, t) = -z\frac{\partial w(x,y,t)}{\partial x}, \\ v(x, y, z, t) = z \cdot \varphi_{yz}(x, y, t) = -z\frac{\partial w(x,y,t)}{\partial y}. \end{cases} \tag{4.28}$$

From the compatibility Eqs. (4.1) the strains are:

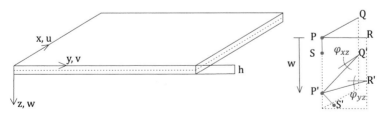

Fig. 4.2 Kinematics of *Kirchhoff–Love* plate under bending. Normals in the deformed configuration continue to be rectilinear, perpendicular to the middle surface and of unaltered length

$$
\begin{cases}
\varepsilon_x(x, y, z, t) = \frac{\partial u(x,y,z,t)}{\partial x} = -z\frac{\partial^2 w(x,y,t)}{\partial x^2}, \\
\varepsilon_y(x, y, z, t) = \frac{\partial v(x,y,z,t)}{\partial y} = -z\frac{\partial^2 w(x,y,t)}{\partial y^2}, \\
\gamma_{xy}(x, y, z, t) = \frac{\partial u(x,y,z,t)}{\partial y} + \frac{\partial v(x,y,z,t)}{\partial x} = -2z\frac{\partial^2 w(x,y,t)}{\partial x \partial y}.
\end{cases}
\tag{4.29}
$$

The in-plane stress state in the layer at distance z can be found immediately by introducing Eq. (4.29) into the constitutive equations reported in Eq. (4.12):

$$
\begin{cases}
\sigma_x(x, y, z, t) = \frac{E}{1-\nu^2}(\varepsilon_x + \nu\varepsilon_y) = -z\frac{E}{1-\nu^2}\left(\frac{\partial^2 w}{\partial x^2} + \nu\frac{\partial^2 w}{\partial y^2}\right), \\
\sigma_y(x, y, z, t) = \frac{E}{1-\nu^2}(\varepsilon_y + \nu\varepsilon_x) = -z\frac{E}{1-\nu^2}\left(\frac{\partial^2 w}{\partial y^2} + \nu\frac{\partial^2 w}{\partial x^2}\right), \\
\tau_{xy}(x, y, z, t) = \frac{E}{2(1+\nu)}\gamma_{xy} = -z\frac{E}{1-\nu^2}(1 - \nu)\frac{\partial^2 w}{\partial x \partial y}.
\end{cases}
\tag{4.30}
$$

Stress resultants (per unit width) are given by bending moments:

$$
\begin{cases}
M_x(x, y, z, t) = \int_{-h/2}^{h/2} \sigma_x(x, y, z, t)z dz = -D\left(\frac{\partial^2 w}{\partial x^2} + \nu\frac{\partial^2 w}{\partial y^2}\right), \\
M_y(x, y, z, t) = \int_{-h/2}^{h/2} \sigma_y(x, y, z, t)z dz = -D\left(\frac{\partial^2 w}{\partial y^2} + \nu\frac{\partial^2 w}{\partial x^2}\right),
\end{cases}
\tag{4.31}
$$

shear forces:

$$
\begin{cases}
T_x(x, y, z, t) = \int_{-h/2}^{h/2} \tau_{xz}(x, y, z, t)dz, \\
T_y(x, y, z, t) = \int_{-h/2}^{h/2} \tau_{yz}(x, y, z, t)dz,
\end{cases}
\tag{4.32}
$$

and twisting moments:

$$
\begin{cases}
M_{xy}(x, y, z, t) = \int_{-h/2}^{h/2} \tau_{xy}(x, y, z, t)z dz = -D(1 - \nu)\frac{\partial^2 w}{\partial x \partial y}, \\
M_{yx}(x, y, z, t) = \int_{-h/2}^{h/2} \tau_{yx}(x, y, z, t)z dz - D(1 - \nu)\frac{\partial^2 w}{\partial x \partial y} = M_{xy},
\end{cases}
\tag{4.33}
$$

where D is the plate bending stiffness, i.e. the flexural rigidity of a strip of unit width, given by:

$$
D = \frac{E}{(1 - \nu^2)}\frac{h^3}{12}
\tag{4.34}
$$

To sum up, we see that the unknowns of the problem are six (w, T_x, T_y, M_x, M_y, M_{xy}), for which we have written three equations relating the bending and twisting moments to w. We need other three equations, which is exactly the number of conditions for the equilibrium of an infinitesimal plate element having sides dx and dy. Writing the moment equilibrium around y and x axis and omitting higher order infinitesimals:

$$
\begin{cases}
T_x(x, y, z, t) = \frac{\partial M_x(x,y,z,t)}{\partial x} + \frac{\partial M_{yx}(x,y,z,t)}{\partial y}, \\
T_y(x, y, z, t) = \frac{\partial M_y(x,y,z,t)}{\partial y} + \frac{\partial M_{xy}(x,y,z,t)}{\partial x}.
\end{cases}
\tag{4.35}
$$

Substituting Eqs. (4.31) and (4.33) into Eq. (4.35) we obtain the expressions of the shear forces T_x and T_y as a function of w:

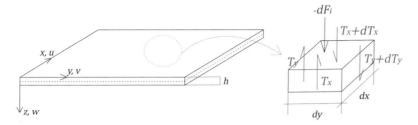

Fig. 4.3 Dynamic vertical equilibrium of *Kirchhoff–Love* plate under shear and inertia forces

$$\begin{cases} T_x(x, y, z, t) = -D\left(\dfrac{\partial^3 w}{\partial x^3} + \dfrac{\partial^3 w}{\partial x \partial y^2}\right), \\ T_y(x, y, z, t) = -D\left(\dfrac{\partial^3 w}{\partial y^3} + \dfrac{\partial^3 w}{\partial x^2 \partial y}\right). \end{cases} \tag{4.36}$$

Finally, we write the dynamic equilibrium equation in the z direction through the vertical equilibrium of an infinitesimal plate element, as in Fig. 4.3:

$$\frac{\partial T_x}{\partial x} dx dy + \frac{\partial T_y}{\partial y} dy dx + \rho \frac{\partial^2 w}{\partial t^2} h dx dy = 0, \tag{4.37}$$

from which:

$$\frac{\partial T_x}{\partial x} + \frac{\partial T_y}{\partial y} + \rho h \frac{\partial^2 w}{\partial t^2} = 0. \tag{4.38}$$

Substituting Eq. (4.36) into Eq. (4.38), we find the governing differential equation for the problem:

$$\frac{\partial^4 w}{\partial x^4} + 2 \frac{\partial^4 w}{\partial x^2 \partial y^2} + \frac{\partial^4 w}{\partial y^4} + \frac{\rho h}{D} \frac{\partial^2 w}{\partial t^2} = 0, \tag{4.39}$$

or, in a more compact notation:

$$\nabla^4 w + \frac{\rho h}{D} \frac{\partial^2 w}{\partial t^2} = 0. \tag{4.40}$$

By considering a harmonic plane wave, as defined in Eq. (3.5), we find the dispersion

$$\omega = \kappa^2 h \sqrt{\frac{E}{12 \rho (1 - v^2)}}, \tag{4.41}$$

that defines the fundamental antisymmetric Lamb mode, usually called A_0 mode (dispersive). Extensional motions are derived from the plane stress governing equations:

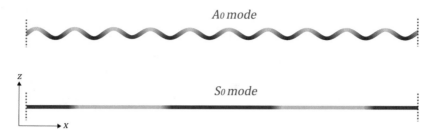

Fig. 4.4 Fundamental antisymmetric (A_0) and symmetric (S_0) modes for a plate computed numerically (FEM)

$$\begin{cases} \frac{\partial \sigma_x}{\partial x} + \frac{\partial \tau_{xy}}{\partial y} - \rho \frac{\partial^2 u}{\partial t^2} = 0, \\ \frac{\partial \tau_{yx}}{\partial x} + \frac{\partial \sigma_y}{\partial y} - \rho \frac{\partial^2 v}{\partial t^2} = 0. \end{cases} \tag{4.42}$$

Substituting the combined constitutive and compatibility equations into the plane stress governing equations (Eq. (4.42))

$$\begin{cases} \sigma_x = \frac{E}{1-v^2} \left(\frac{\partial u}{\partial x} + v \frac{\partial v}{\partial y} \right), \\ \tau_{xy} = \tau_{yx} = \frac{E}{2(1+v)} \left(\frac{\partial u}{\partial y} + \frac{\partial v}{\partial x} \right), \end{cases} \tag{4.43}$$

we obtain:

$$\begin{cases} \frac{\partial^2 u}{\partial x^2} + \frac{1-v}{2} \frac{\partial^2 u}{\partial y^2} + \frac{1+v}{2} \frac{\partial^2 v}{\partial x \partial y} = \frac{(1-v^2)\rho}{E} \frac{\partial^2 u}{\partial t^2}, \\ \frac{\partial^2 v}{\partial y^2} + \frac{1-v}{2} \frac{\partial^2 v}{\partial x^2} + \frac{1+v}{2} \frac{\partial^2 u}{\partial x \partial y} = \frac{(1-v^2)\rho}{E} \frac{\partial^2 v}{\partial t^2}. \end{cases} \tag{4.44}$$

Substituting again the harmonic plane wave solution defined by Eq. (3.5), we obtain

$$\omega = \sqrt{\frac{E}{\rho(1-v^2)}} \kappa \tag{4.45}$$

that defines the fundamental symmetric Lamb mode, usually called S_0 mode (non-dispersive). We report in Fig. 4.4 the A_0 and S_0 Lamb modes obtained from a 2D numerical (FEM) analysis in Abaqus with infinite elements at the plate edges.

By looking at the ratio between the dispersive A_0 and non-dispersive S_0 Lamb modes (Eqs. (4.41) and (4.45)) for wavelength/thickness ratios that satisfy the Kirchhoff–Love theory, this is always lower than one. This implies that, for a given frequency, the wavelength associated to the S_0 mode is higher than the one of the A_0 mode.

4.3 Wave Propagation in Thin Elastic Plates with Resonators

We now investigate the influence of resonators introduced on an elastic plate. An analytical detailed analysis is reported in [3], while here we show numerical results obtained through the FEM software Abaqus.

A cluster of rods attached to a plate can act, for short pillars, as a phononic crystal [4, 5], or a resonant metamaterial for long rods [6]. This system, endowed with non-conventional dispersion properties, is a metamaterial, and, more specifically, a *metasurface* [7] due to its interaction with surface waves. Among the possible resonant metasurface designs for elastic waves recently proposed [8–13], the one made of a cluster of rods (the resonators) [4, 5, 14] on an elastic substrate has revealed superior characteristics and versatility of use. A single rod attached to an elastic plate couples with the motion of the A_0 mode. This coupling is particularly strong at the longitudinal resonance frequencies of the rod [3, 15, 16]. The impact of the flexural resonances on the wave propagation is negligible [17], at least in terms of the band-gaps. At this point, the eigenvalues of the equation describing the motion of the substrate and the rod are perturbed by the resonance and become complex leading to the formation of a band-gap [18, 19]. Due to the dominant contribution of the longitudinal resonance, we normalize the results with respect to the first longitudinal (axial) natural frequency of a clamped beam:

$$\omega_0 = \frac{\pi}{2h}\sqrt{\frac{E}{\rho}}. \tag{4.46}$$

This solution can be obtained solving, with the separation of variables method the d'Alembert equation reported in Eq. (3.4) and imposing fixed-free boundary conditions. More immediately, it is possible to obtain Eq. (4.46) from (3.6) introducing the wavelength $\lambda = 4h$, i.e. $\kappa = \pi/2h$.

Equation (4.46) holds provided the substrate is sufficiently firm so that one end of the rod is effectively clamped. Conversely, an overly soft substrate (e.g. a very thin plate) may result in a rod behaving as a rigid body connected with a spring. In practical terms, this provides bounds on the plate thickness and requires cross-checking the modal shape of the longitudinal eigenmode. When the resonators are arranged on a subwavelength cluster, the resonance of each rod acts constructively enlarging the band-gap until, approximately, the rod's anti-resonance [6, 15]. In Fig. 4.5, we show the 2D numerical dispersion curves of a plate with a periodic arrangement of resonators. Specifically we consider two plate thickness values: $t_1 = h/30$ and $t_2 = h/15$. It can be noticed that increasing the plate thickness, the accuracy of Eq. (4.46) increases.

On the other hand, we see that the phenomenon changes completely when the thickness of the plate is very small, or the height of the rod is reduced (see Fig. 4.6). In this case, the behaviour is more like a phononic crystal.

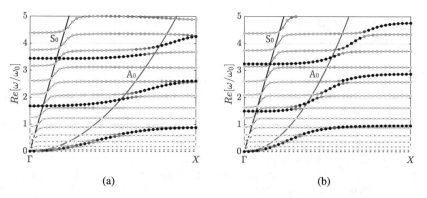

(a)   (b)

Fig. 4.5 Numerical (FEM) dispersion curves for a plate with resonators ($a = 30$ mm, $h = 300$ mm), with thickness $t_1 = h/30$ (**a**) and $t_2 = h/15$ (**b**). The solution ω converges to ω_0 by increasing the plate thickness. Coloured dots represent the wave polarization in the resonator, with vertical (black) and horizontal (white) motion

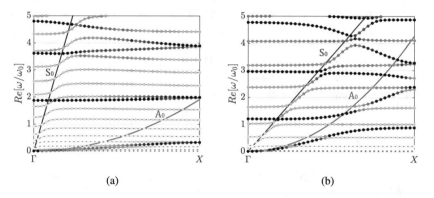

(a)   (b)

Fig. 4.6 Numerical dispersion curves for a plate with resonators ($a = 30$ mm, $\bar{h} = h/3 = 100$ mm) with thickness $t_1 = \bar{h}/100$ (**a**) and $t_2 = \bar{h}/5$ (**b**). Coloured dots represent the wave polarization in the resonator, with vertical (black) and horizontal (white) motion

It is important to notice that the array interact properly with the A_0 mode only. This can be clearly seen exciting the metasurface with an A_0 or S_0 mode, as shown in Fig. 4.7. The different behaviour is due to the strong mismatch between the axial stiffness of the plate and the flexural one of the resonators, as well as the higher wavelength of the S_0 mode with respect to A_0. Indeed, by fixing a certain frequency, the wavenumber associated to S_0 is lower with respect to A_0 (see for e.g. Fig. 4.5).

Graded plate. Because the resonance frequency of the rod, which depends on rod height only, determines the band-gap position, then by simply varying the length of the rods one gets an effective band-gap that is both broad and subwavelength; the rod length is therefore a key parameter for the metasurface tunability. This is a key concept in what we define as a *graded* system, i.e. a medium characterised by a gentle spatial variation of the effective properties. Due to the adiabatic changes of

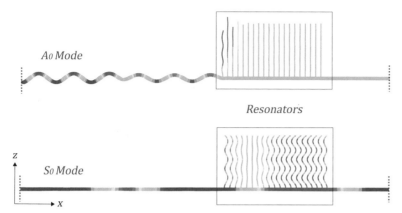

Fig. 4.7 Elastic wave propagation in a plate with a cluster of resonators, excited with the A_0 or S_0 mode at the same frequency. While the resonators interacts with the A_0 mode, creating a band gap, this is not true for S_0, mainly due to it's higher wavelength for the same frequency of excitation

the array parameters, the global spatial properties of the array are determined by the locally periodic structures; the dispersive properties of the array at a given position are inferred from the periodic dispersion curves of the constituent element at that position [20–22]. Due to the capability to spatially separate the signal depending on frequency, this effect is called *rainbow*, as firstly analysed in electromagnetism [23] and then in acoustics [24].

Specifically, by exciting the system from short to high resonators, we see that the waves slow down until the band gap opening, while in the opposite direction the waves are immediately stopped and reflected backward (see Fig. 4.8). Slowed waves are associated to a group velocity reduction and increase in the wavenumber, i.e. wavelength reduction. This results in an amplification of the wavefield in the resonators due to the longer interaction with the waves. In addition, it is worth to notice that different modes are involved: in the rainbow case, we have a behaviour dominated by longitudinal resonances, while in the opposite direction by flexural modes.

4.4   Wave Propagation in Unbounded Media and Half-Spaces

Let us now consider a plane wave propagating with phase velocity c:

$$\mathbf{u} = f(\mathbf{x} \cdot \mathbf{p} - ct)\mathbf{d}, \tag{4.47}$$

where $\mathbf{p}$ and $\mathbf{d}$ are unit vectors defining the direction of propagation and motion respectively. The vector $\mathbf{x}$ denotes the position vector, and $\mathbf{x} \cdot \mathbf{p} = constant$

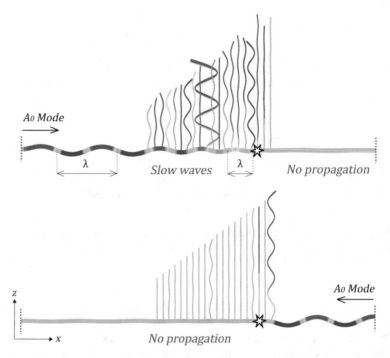

Fig. 4.8 Wave propagation in a plate with a graded array of resonators. When the A_0 mode propagates from short to high resonators, the wave is slowed down and the wavelength is reduced. For propagation in the opposite direction (at the same frequency), a band gap suddenly opens in the position of the first resonator, and the wave is immediately reflected backward

describes a plane normal to the unit vector p. The Eq. (4.47) represents a plane wave, whose planes of constant phase are normal to p and propagate with velocity c. Substituting the plane wave expression Eq. (4.47) into the field Eqs. (4.18), and adopting these relations

$$\begin{cases} \nabla \cdot \boldsymbol{u} = (\boldsymbol{p} \cdot \boldsymbol{d}) f'(\boldsymbol{x} \cdot \boldsymbol{p} - ct), \\ \nabla \nabla \cdot \boldsymbol{u} = (\boldsymbol{p} \cdot \boldsymbol{d}) f''(\boldsymbol{x} \cdot \boldsymbol{p} - ct) \boldsymbol{p}, \\ \nabla^2 \boldsymbol{u} = f''(\boldsymbol{x} \cdot \boldsymbol{p} - ct) \boldsymbol{d}, \\ \ddot{\boldsymbol{u}} = c^2 f(\boldsymbol{x} \cdot \boldsymbol{p} - ct) \boldsymbol{d}, \end{cases} \qquad (4.48)$$

we obtain:

$$(\mu - \rho c^2) \boldsymbol{d} + (\lambda + \mu)(\boldsymbol{p} \cdot \boldsymbol{d}) \boldsymbol{p} = 0. \qquad (4.49)$$

Since p and d are two different unit vectors, Eq. (4.49) can be satisfied if:

- either $\boldsymbol{d} = \pm \boldsymbol{p}$, or $\boldsymbol{p} \cdot \boldsymbol{d} = 0$:

$$c = c_L = \sqrt{\frac{\lambda + 2\mu}{\rho}}. \tag{4.50}$$

In this case the motion is parallel to the direction of propagation and the wave is therefore called a *longitudinal* wave. By computing the rotation components we get:

$$\nabla \wedge \boldsymbol{u} = (\boldsymbol{p} \wedge \boldsymbol{d}) f'(\boldsymbol{x} \cdot \boldsymbol{p} - ct) = 0. \tag{4.51}$$

The rotation thus vanishes, which has motivated the alternative terminology of *irrotational* wave. This type of wave is usually called *dilatational* wave, *pressure* wave or P-wave (Primary, Pressure).

- both terms vanish independently in the case $\boldsymbol{d} \neq \pm\boldsymbol{p}$:

$$\boldsymbol{p} \cdot \boldsymbol{d} = 0 \text{ and } c = c_T = \sqrt{\frac{\mu}{\rho}}. \tag{4.52}$$

Now the motion is normal to the direction of propagation, and the wave is called a *transverse* wave. The divergence of the displacement vector gives:

$$\nabla \cdot \boldsymbol{u} = (\boldsymbol{p} \cdot \boldsymbol{d}) f'(\boldsymbol{x} \cdot \boldsymbol{p} - ct) = 0, \tag{4.53}$$

and then the wave is equivoluminal. This type of wave is usually called a *rotational* wave, *shear* wave or S-wave (Secondary, Shear).

For the special case of harmonic plane waves of the form

$$\boldsymbol{u}(\boldsymbol{x}, t) = A\boldsymbol{d}e^{i\kappa(\boldsymbol{x} \cdot \boldsymbol{p} - ct)}, \tag{4.54}$$

we have two types of plane harmonic waves propagating with phase velocities c_L and c_P. Since the wavenumber κ does not appear in the expressions for the phase velocities, the dispersion relation is linear and then plane harmonic waves in an unbounded homogeneous, isotropic, linearly elastic medium are non dispersive.

Considering a half-space, the wave can travel along the free surface, as firstly considered by Lord Rayleigh [25]. The criterion for surface, or *Rayleigh* (R), waves is that the displacement decays exponentially with distance from the free surface (i.e. we enforce it to be on the surface only). We investigate the Rayleigh wave propagation for the two-dimensional case of plane waves propagating along a given direction x_1, as shown in the schematic of Fig. 4.9.

Let us consider displacement components of the form:

$$\begin{cases} u_1(x_1, x_2, t) = Ae^{i\kappa(x_1-ct)-bx_2}, \\ u_2(x_1, x_2, t) = Be^{i\kappa(x_1-ct)-bx_2}, \\ u_3(x_1, x_2, t) \equiv 0. \end{cases} \tag{4.55}$$

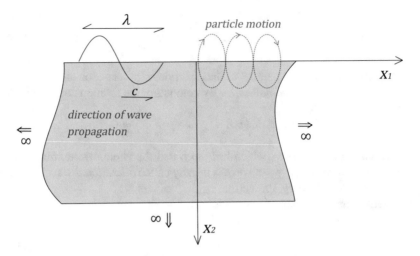

Fig. 4.9 Schematic of surface, or Rayleigh, wave propagation in the free surface of an elastic half-space. The wavefield decays exponentially with distance from the free surface

The real part of b is positive, so that the displacement decreases for increasing x_2. i.e. the depth of the space. By substituting Eq. (4.55) inside the equation of motion Eq. (4.18) two homogeneous equations for the constants A and B are obtained. A nontrivial solution exists if the determinant of the coefficients vanishes, i.e.

$$[c_L^2 b^2 - (c_L^2 - c^2)\kappa^2][c_T^2 b^2 - (c_T^2 - c^2)\kappa^2] = 0, \tag{4.56}$$

whose roots are:

$$b_1 = \kappa\sqrt{1 - \frac{c^2}{c_L^2}}, \quad b_2 = \kappa\sqrt{1 - \frac{c^2}{c_T^2}}. \tag{4.57}$$

In order to have real valued solutions: $c < c_T < c_L$. The ratios corresponding to b_1 and b_2 are then given by:

$$\left(\frac{B}{A}\right)_1 = -\frac{b_1}{i\kappa}, \quad \left(\frac{B}{A}\right)_2 = \frac{i\kappa}{b_2}. \tag{4.58}$$

Returning to Eq. (4.55), a general solution of the displacement equations of motion may thus be written as:

$$\begin{cases} u_1(x_1, x_2, t) = [A_1 e^{-b_1 x_2} + A_2 e^{-b_2 x_2}]e^{i\kappa(x_1 - ct)}, \\ u_2(x_1, x_2, t) = [-\frac{b_1}{i\kappa} A_1 e^{-b_1 x_2} + \frac{i\kappa}{b_2} A_2 e^{-b_2 x_2}]e^{i\kappa(x_1 - ct)}, \\ u_3(x_1, x_2, t) \equiv 0. \end{cases} \tag{4.59}$$

The constants A_1 and A_2 and the phase velocity c have to be choosen such that the stresses σ_{22} and σ_{21} vanish at $x_2 = 0$, i.e. the free surface. By substituting Eq. (4.59) into the expressions of σ_{22} and σ_{21} at $x_2 = 0$ we get:

$$\begin{cases} 2b_1 A_1 + \left(2 - \frac{c^2}{c_T^2}\right) \kappa^2 \frac{A_2}{b_2} = 0, \\ \left(2 - \frac{c^2}{c_T^2}\right) A_1 + 2b_2 \frac{A_2}{b_2} = 0. \end{cases} \tag{4.60}$$

A nontrivial solution exists if the determinant of the coefficients vanishes, i.e.

$$\left(2 - \frac{c^2}{c_T^2}\right)^2 - 4\sqrt{\left(1 - \frac{c^2}{c_L^2}\right)\left(1 - \frac{c^2}{c_T^2}\right)} = 0. \tag{4.61}$$

Since the wavenumber does not enter in Eq. (4.61), Rayleigh surface waves at a free edge of an elastic half-space are non dispersive. A good approximation, for linear elastic materials with positive Poisson ratio, of the Rayleigh phase velocity c_R can be written as [1]:

$$c_R = \frac{0.862 + 1.14\nu}{1 + \nu} c_T. \tag{4.62}$$

By looking at the ratio u_2/u_1:

$$\frac{u_2}{u_1} = e^{i\frac{\pi}{2}} \left[\frac{\frac{b_1}{\kappa} A_1 e^{-b_1 x_2} + \frac{\kappa}{b_2} A_2 e^{-b_2 x_2}}{A_1 e^{-b_1 x_2} + A_2 e^{-b_2 x_2}} \right], \tag{4.63}$$

it is clear that the displacement components are $\pi/2$ out of phase, and then the trajectories of the particles are ellipses, as shown in Fig. 4.9. In Fig. 4.10 we visualize P, S and Rayleigh (R) waves from a 2D numerical analysis performed using the FEM commercial software Abaqus. The wavefield is obtained imposing a vertical force $F(t)$ able to simultaneously generate bulk (P and S) and surface (R) waves. The boundary of the half-space is modeled using infinite elements, in order to avoid disturbances due to spurious reflections. The nature of the propagating waves can be distinguished by looking at the different motion of the particles: parallel to the direction of propagation for P, normal for S, and elliptical trajectories for R. Moreover, each wave is associated (for the same frequency of excitation) to a different wavelength and phase velocity.

4.5 Wave Propagation in Half-Spaces with Resonators

We now investigate the influence of resonators introduced on an elastic half-space. We report here numerical (FEM) analyses performed in Abaqus, while a detailed analytical model can be found in [3]. The introduction of resonators, i.e. locally resonant

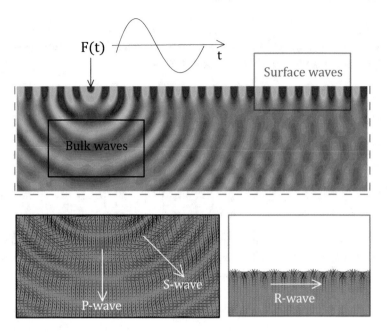

Fig. 4.10 Wave propagation in an elastic half-space with a detailed view of the wave fluxes for both bulk and surface waves

structures, modifies the medium dispersion properties. In this way, Rayleigh waves can be surprisingly manipulated, with the creation of band gaps or mode conversions, as firstly demonstrated for seismic waves in [15]. The obtained system is really a metamaterial, due to it's non-conventional dispersion properties, e.g. negative refraction. As already said, metamaterials should be distinguished from phononic crystals that typically show Bragg-scattering induced band gaps [26, 27], with wave manipulation at the wavelength scale. On the contrary, the sub-wavelength arrangement of resonators [28–31] allows control of waves at a deep sub-wavelength scale (approximately up to $\lambda/15$). Because the resonators are arranged on a sub-wavelength scale, the cumulative effect of several resonators over a wavelength interferes constructively [32] thus creating a band gap between the resonance and anti-resonance. More precisely, the proposed system is a *metasurface* [7], capable of manipulating the propagation of surface waves, while metamaterials were initially developed for bulk media.

We start looking at the dispersion curves for a unit cell of infinite depth (modeled in Abaqus using infinite elements) with a resonator on the top (see Fig. 4.11). In this physical model, the role played by the flexural resonances is marginal, due to the Rayleigh wave polarization that predominantly excite the vertical direction. We consider two cases with equal unit cell size $a = 30$ mm, but different resonator's height. Specifically, we consider $h_1 = 2a$ and $h_2 = 5a$. We see that in both cases (see Fig. 4.11a, b) due to the resonators, the problem becomes dispersive, with zero

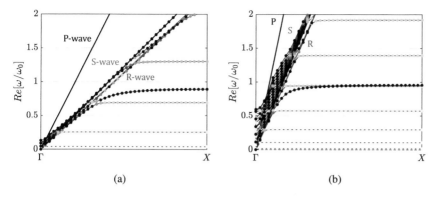

Fig. 4.11 Numerical (FEM) dispersion relation for a unit cell composed of a half-space cell with a beam on the free surface. Results are normalized with respect to ω_0 and two values of beam height are considered: **a** $h_1 = 2a$ and **b** $h_2 = 5a$ where $a = 30$ mm is the unit cell dimension. Colours represent the wave polarization inside the beam: horizontal (white) or vertical (black), corresponding to flexural and axial motion respectively

group velocity modes reached at the edge of the Brillouin Zone (point X). These waves arise as a result of the periodicity of the array present on the boundary of the half-space and are called *Rayleigh–Bloch* waves [33, 34]. Since the wave is assumed to propagate from $\Gamma \rightarrow X$, and we can in principle move only on R, S or P dispersion lines for a plain half-space, the curves above P are spurious unphysical solutions [35] found by the eigenvalue solver; these arise from the finite depth of the simulated region and correspond to propagating modes or modes created by the finite layer.

By comparing the two cases reported in Fig. 4.11a, b, we can see that increasing the height of the beam, the solution converges at the X point, i.e. where the wave is reflected backward, to the axial resonance of a clamped beam, as identified form Eq. (4.46). Indeed, increasing the height of the beam, the axial resonance appears at a lower frequency and higher wavelength, reducing the influence of the boundary conditions (the base of the beam moves without deforming, i.e. like a clamp). On the other hand, by decreasing the height of the beam, the system start behaving more like a phononic crystal. This transition between local resonance and Bragg can be noticed by looking at the shape of the dispersion curve corresponding to the longitudinal motion of the beam. In a local resonance dominated response, the dispersion curve remains flat over a large $\Gamma - X$ region (suggesting interaction with low wavenumbers), while in a Bragg dominated response the zero group velocity is reached almost only at the X point.

We now look at the wavefield for a frequency below and above the band gap opened by the axial resonance. Following the lower dispersion curve branch, we see a reduction of the wave group velocity (slow Rayleigh waves), until the band gap opening (Fig. 4.12). We also note the decrease in wavelength inside the array corresponding to the slower effective wavespeed inside the metasurface as compared with the homogeneous half-space. Moving on to Fig. 4.13, we notice the upper branch

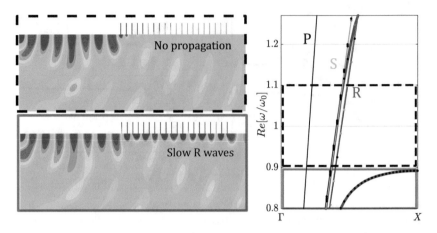

Fig. 4.12 Wave propagation in an elastic half-space with resonators. Following the lower dispersion branch, Rayleigh waves are slowed down, until the band gap opening

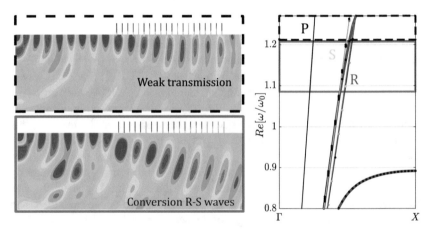

Fig. 4.13 Wave propagation in an elastic half-space with resonators. Following the higher dispersion branch, Rayleigh waves are mode converted into S-waves

transition from Rayleigh to the shear waves, that is, the Rayleigh–Bloch wave in the resonant array preserves the Rayleigh-like surface wave properties at higher frequencies but will evolve into a shear-like wave as the frequency decreases.

Graded half-space. Adopting graded designs, i.e. gradually varying the rods' resonance frequencies along space, we get the so called *metawedge* [22] that can either mode convert incident surface Rayleigh waves into bulk elastic shear waves, or reflect the Rayleigh waves creating an elastic rainbow, analogous to the optical rainbow for microwaves [23]. A key concept of the optical rainbow effect [23] is the graded wedge waveguide with a negative-index left-handed core that can trap and spatially segregate (as a rainbow) the different frequencies (colours) of light. Similarly to this

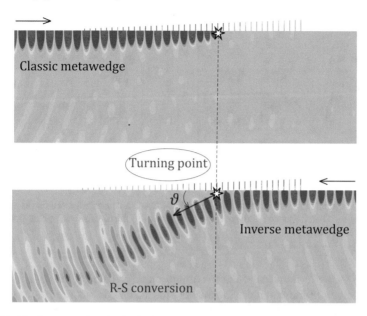

Fig. 4.14 Elastic metawedge obtained with a graded array of resonators on an elastic half-space. Rainbow effect is obtained with the classic metawedge, while $R - S$ conversion with the inverse metawedge. The observed turning point in both cases, remains approximately the same

in elasticity [22, 36], a graded array of vertical resonator can spatially stop (*classic metawedge*) or mode convert (*inverse metawedge*) Rayleigh waves, depending on frequency. For the classic wedge case, the incident wavefront approaches the wedge from the left towards the short edge of the metawedge. After travelling a few wavelengths inside the wedge, it slows down until reaching the resonator whose fundamental longitudinal mode matches the input frequency of the signal. In this position, called *turning point*, a band gap opens and the energy is reflected backward. The wavefront initially encounters very short resonators and it propagates as a Rayleigh mode. As the height of the resonators increases, the group velocity is progressively reduced until the band gap opening. On the other hand, for the inverse metawedge case, at the turning point the wavefront is mode-converted and directed into the interior of the substrate with the Rayleigh waves, characterized by x and z-components, converted into a S-wave whose motion is polarized in the transverse direction.

The converted wave angle shown in Fig. 4.14 and obtained through FEM simulations in Abaqus, can be well predicted by the Snell's law:

$$\theta = \arccos \left(\frac{c_R}{c_T} \right). \tag{4.64}$$

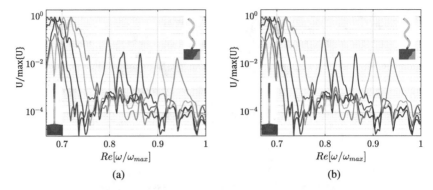

Fig. 4.15 Normalized strain energy density in the last five resonators in the classic (**a**) and inverse (**b**) metawedge. The frequency is normalized with respect to the highest operational frequency of the metawedge ω_{max}

It is interesting to notice that the maximum energy taken by the rods in the two cases involves different modes. Specifically, for the classic metawedge we reach the maximum kinetic/strain energy for the axial modes, while for the inverse metawedge for the flexural one. This can be noticed looking at the normalized strain energies in the resonators for the classic and inverse metawedge (Fig. 4.15). Considering for e.g. the last five resonators, it can be noticed that in the metawedge operational frequency range (from ω_{min} to ω_{max}) the amount of energy taken by the vibration modes changes from the conventional to the inverse metawedge, as shown in Fig. 4.15. This is due to the different way in which the resonators are excited: while in the rainbow effect the wave is slowed down, providing mainly a vertical excitation, in the conversion mechanism the excitation is more like the isolated beam case, with strong rotational components.

4.6   Reversed Mode Conversion Using Umklapp

It is interesting to notice that a half-space with a graded array can exploit other conversion mechanisms going outside of the first BZ, as explained in [37]. Specifically, one can obtain *reversed* mode conversion from surface (R) waves to compressional P-waves and shear S-waves in the bulk. This mechanism can be explained through a phenomenon known in physics as *Umklapp*. By assuming a wavevector beyond the first BZ, due to periodicity properties (see for instance Fig. 3.8), it is possible to represent it inside the FBZ, by a translation of $\kappa d = -2\pi i$. In terms of wavevectors, a translation of $\mathbf{G} = (-2\pi/d)\mathbf{i}$ is needed. This is called *reciprocal vector*, since belongs to the *reciprocal* space, i.e. the reciprocal of the physical space: $d \mapsto 1/d$. Rigorous demonstrations can be found in famous solid state physics books [38–41].

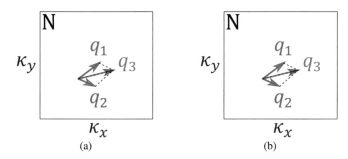

Fig. 4.16 Wave vector scattering in the first BZ showing conventional definitions of a N-process (**a**) and a U-process (**b**)

Historically, the Umklapp, flip-over, or U-process, was firstly hypothesised by Peierls [39] in describing phonon-phonon (i.e. the quantum mechanical description of an elementary vibrational motion) scattering to explain thermal conductivity at high temperatures, and has a rich history in the quantum theory of thermal transport [42]. The conventional definition of a U-process is described in Fig. 4.16b, whereby the resultant wavevector of a scattering process within a periodic crystal lies outside the first BZ. If the sum of q_1 and q_2 stay inside the first BZ (black square), the process is called normal scattering (N-process). If the sum of q_1 and q_2 defines a wavevector outside the first BZ, this vector can be equivalently represented inside the first BZ by the addition of a reciprocal lattice vector $\mathbf{G}$. Using conventional textbook distinction [39]:

$$\mathbf{q}_1 + \mathbf{q}_2 - \mathbf{q}_3 = \begin{cases} \mathbf{0} & \text{N-process,} \\ \mathbf{G} & \text{U-process.} \end{cases} \tag{4.65}$$

where the $\mathbf{q}_i$ are the wave vectors in Fig. 4.16. Reversed mode conversion can be achieved by utilising the existence surface waves outwith the periodic structure, and marrying the transition of such a wave to the excited wave within a graded structuring. This is well predicted by looking at the dispersion curves of the locally periodic elements, along with simplified isofrequency contours. We consider an array composed of rods with diameter $d = 0.5$ mm, periodicity $a = 1.5$ mm, initial height $h_0 = 0.5$ mm and grading $\Delta h = 0.05$ mm. In Fig. 4.17a the dispersion curves for the longitudinal motion of rods within the second BZ are reported. These curves are obtained through an averaging process of the fully polarised dispersion curves. The relative degree of longitudinal/flexural motion for higher modes is not as clear as for the lowest modes. To extract the excited, localised surface modes, a weighted averaging technique is used to interpolate, and hence separate, both motions independently. The curves with the opposite behaviour below a set tolerance on the vertical and horizontal displacements of the rods are removed.

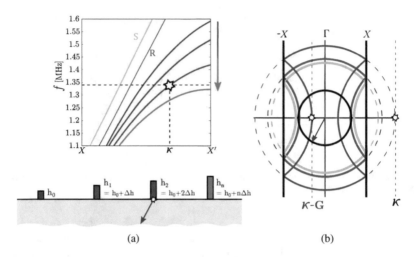

(a)   (b)

Fig. 4.17 **a** Longitudinal dispersion curves within second BZ; X marks the edge of the first BZ, whilst X' marks the edge of the second BZ. Each dispersion curve represents the second longitudinal mode for a perfectly periodic array of rods. **b** Isofrequency contours of rod where last longitudinal mode supported, after which an effective band gap is reached, when exciting at a frequency of 1.340 MHz. At this position U-processes are preferential to a change in rod motion and, resulting in an effective reversed wavenumber $\kappa - G$, capable of mode converting into a P-wave (black isofrequency contour) [37]

By increasing the rod height, an effective band gap opens in a κ position where U-processes dominate; The transfer of crystal momentum results in an effective reversed wavenumber $\kappa - G$. Depending on the operational frequency, this can lie within the isocircle of the free S or P body waves; reversed conversion is achieved by conservation of the tangential component of the wavevector, as shown in Fig. 4.17b. The isofrequency contour is a superposition of the projections of the simplified contours of the system into one plane; despite being a 1D periodic system, we show isocircles to aid visualization and prediction of the angles of the reversed conversion. Figure 4.18 shows a detailed description of this procedure. To determine the angles and polarisation of the reversed conversion effect we analyse the incident wavevector, and how it is altered, relative to the first BZ of the periodic array, for the dispersion curves presented in Fig. 4.17a. We draw the 1D first BZ of the periodic array as an infinite strip, allowing us to relate the contours of the bulk and surface waves of the free material to the array. Since the Rayleigh wave is excited in the free, isotropic surface this is indeed an isocircle, as shown in Fig. 4.18a. By grading the array, we increase the wavenumber until the band gap is reached (see Fig. 4.18b, c) whose colors are related to the dispersion curves in Fig. 4.17a. Since the wavevector is outside the first BZ, we have to apply a reciprocal vector translation, obtaining reverse conversion (Fig. 4.18e, f). By exploiting this mechanism, striking reversed conversion into S and P waves can be achieved, and controlled, as predicted by numerical (FEM) simulations performed using Specfem and Abaqus (Fig. 4.19).

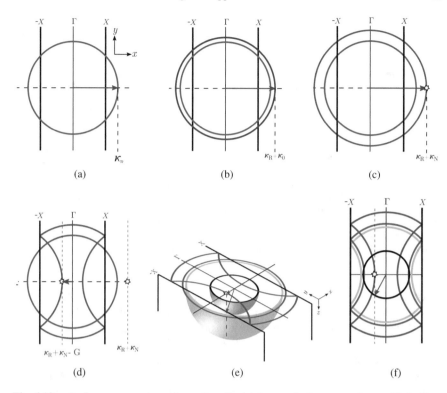

Fig. 4.18 **a** Isofrequency contour of a surface Rayleigh wave in the $x - y$ plane, with incident Rayleigh wavenumber, κ_R, marked by the blue arrow. **b** Isofrequency contour for the first encountered rod: the incident Rayleigh wave excites a guided wave with wavevector $\kappa_R + \kappa_0$; **c** Last supported wavevector, at the Nth rod, $\kappa_R + \kappa_N$. **d** Translated, or flipped, wavevector by a reciprocal lattice vector $\mathbf{G} = (2\pi/a)\mathbf{i}$, with a the periodicity. **e** Isosurfaces of the body waves in the z direction, relative to the first BZ. These are half-spheres as the bulk material is homogeneous. By conserving the tangential component of the reversed wavevector in this 3D picture we can see which, if any, of the shear and compressional isosurfaces are intersected. This gives us the prediction and polarisation of the reversed body wave. **f** Compactified version of **e**, projecting all contours into the same plane [37]

It is important to notice that Umklapp scattering occurs independently of any grading parameters, so long as the frequencies of excitation result in wavevectors that are outwith the first BZ. For this case, a mixture of both S and P waves is observed, since there is no effective band gap to confine the conversion into a confined beam of a single wave type. The amplitude of the localised wave is reduced as it transits the array as a result of crystal momentum transfer at every point along the array and this then forms a leaky elastic antenna waveguide [37].

We notice that it is possible to combine Umklapp phenomenon with *self-phasing* [35]. The self-phasing effect arises due to the propagation of the surface wave through the graded region until the effective band gap is reached and the surface wave then slows down and stops. Once the phase has been altered, such that, in reciprocal

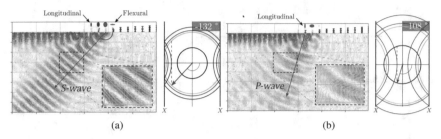

(a)   (b)

Fig. 4.19 Simulations of reversed conversion (using Specfem) by Umklapp scattering: **a** S-conversion for excitation at $f = 1.20$ MHz, and **b** P-conversion for excitation at $f = 1.45$ MHz. Ellipses above the rods show the ratio of the relative magnitude of longitudinal to flexural motion. The angle of conversion, computed directly from the simulated field transformed in the wavenumber space, matches the predictions from the isofrequency curves of the last rods supporting the longitudinal mode. The difference between the converted wave types (S and P) can be seen clearly from the streamline analysis. Wavenumber analysis on the simulated data also confirmed the two different velocities approximately 3000 m/s for S and 6000 m/s for P [37]

space, the frequency and wavenumber are at the band-edge reflection then occurs; upon reflection, the wave is endowed with a phase change of $-\pi/a$, with a being the periodicity. The self-phasing effect is intrinsically different to the use of Umklapp scattering; for U-processes we require the transfer of crystal momentum, whereas for passive self-phased arrays we utilise reflection at the band edge [37].

These conversion mechanisms can be adopted to localize energy from the surface, to specific positions in the bulk (Fig. 4.20). This can be done for e.g. by defining two arrays, on the left and the right of the source (similarly in 3D with circles). Depending on the frequency of excitation, the focus position changes, as shown in Fig. 4.20a–c through Abaqus simulations.

Experiments on Umklapp conversion. The proposed numerical and theoretical results are confirmed experimentally in a 1.8 cm thick slab of aluminum patterned using 3D printing, with a graded array of aluminum microrods on the surface (Fig. 4.21). The aluminium resonators are printed by selective laser melting (SLM).

The block is attached to a moving platform. A laser adaptive photorefractive interferometer scans the surface of the aluminium sample providing a reading of the displacement field in the out of plane direction u_z. Pure Rayleigh waves are generated by an ultrasonic transducer attached at the surface of the aluminium sample. The Rayleigh wave is generated by a Panametrics Videoscan V414 0.5 MHz plane wave transducer and coupled into the sample by a 65° polymer wedge. Phenyl salicylate was used to glue the transducer and wedge to the sample providing good coupling and long term stability. A Ritec $RPR - 4000$ programmable pulser drove the transducer using a 3-cycle sinusoidal burst at 1 and 1.5 MHz for S and P-conversion, respectively, with an amplitude of 300 V peak-to-peak and repetition rate of 500 Hz. At this repetition rate, there were no echoes from previous pulses. The sample was mounted on scanning stages and measured with a rough-surface capable optical detector (Bossanova Tempo-10HF) over an area of 100×30 mm^2 and we used a

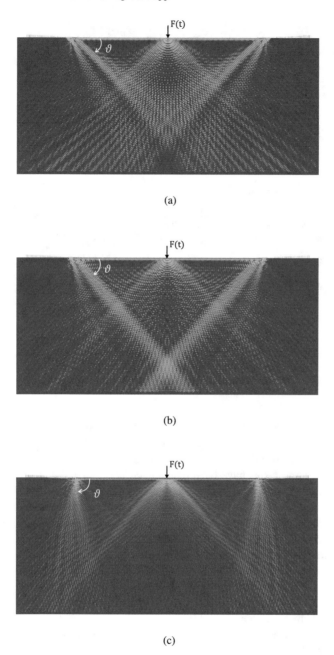

(a)

(b)

(c)

Fig. 4.20 Bulk waves focusing from conversion of surface Rayleigh waves. Depending on frequency the orientation changes, i.e. the angle θ: **a** 1.01 MHz, **b** 1.41 MHz, and **c** 1.75 MHz respectively

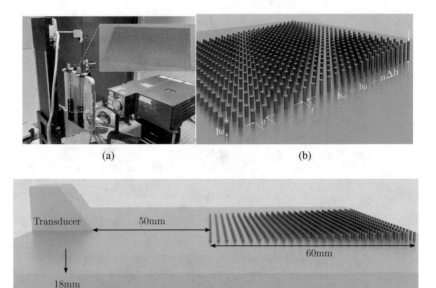

Fig. 4.21 **a** Experimental setup and **b** schematic of array geometry detailing the rod diameter, t, periodicity a and grading through the changes in height of the nth rod, h_n, as $h_n = h_0 + n\Delta h$. **c** Schematic of the arrangement between the transducer and the array, with the plate width [37]

0.25 mm step-size. An Agilent digital oscilloscope was used to captured the signal with 125 MSa/s and 512 averages before transfer to a desktop computer.

To enhance the visualisation of the conversion spatial and frequency filters have been applied. Time series have been bandpassed between 1.1–1.2 MHz and 1.45–1.55 MHz for S and P conversion respectively. The wavefield scans have been filtered selecting wavevectors pointing towards (away) from the resonators for the top (bottom) surface. This procedure mainly remove echoes from the boundaries as well as leakage from the transducer. The wavelength difference between input Rayleigh and converted S-waves suggests that the wavefront is mainly reflected upward in the bulk according to Snell law only partly converting into backward travelling Rayleigh waves along the bottom surface. In the P conversion case, the effect is exacerbated and differences in wavelength, velocity and propagation direction between top and bottom signal are clear (see Fig. 4.22).

Reversing the orientation of the array, i.e. going from tall to short resonators, the reversed Umklapp conversion does not results in a well confined beam; the array behaves as a leaky antenna waveguide, as shown in [37].

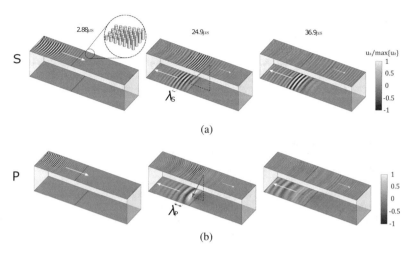

Fig. 4.22 Experimental snapshots in time of temporal-spatially filtered scans along top and bottom surfaces observing S conversion (**a**) and P conversion (**b**), filtered between 1.1–1.2 MHz and 1.45–1.55 MHz respectively and normalised to the maximum of displacement of the top surface (u_z). Solid red lines show where graded array begins, with increasing rod height in the direction of wave propagation on the top surface. The reversed conversion is clear; the forward propagating surface wave on the top surface is reverse converted into the bulk and is seen to excite reversed propagating surface waves on the bottom surface. The measured angles of reversed conversion, $-131.8°$ and $-106.9°$ for S and P respectively, match the predicted theoretical angles, with the separate polarisations evident from the difference in excitation wavelength on the bottom surface, marked λ_S and λ_P [37]

References

1. J.D. Achenbach, S.A. Thau, Wave propagation in elastic solids. J. Appl. Mech. (1974)
2. A.E.H. Love, *Treatise on Mathematical Theory of Elasticity*. Dover Books on Engineering, 4th edn. (1944)
3. D.J. Colquitt, A. Colombi, R.V. Craster, P. Roux, S.R.L. Guenneau, Seismic metasurfaces: sub-wavelength resonators and Rayleigh wave interaction. J. Mech. Phys. Solids (2017)
4. T.T. Wu, Z.G. Huang, T.C. Tsai, T.C. Wu, Evidence of complete band gap and resonances in a plate with periodic stubbed surface. Appl. Phys. Lett. (2008)
5. Y. Achaoui, A. Khelif, S. Benchabane, L. Robert, V. Laude, Experimental observation of locally-resonant and Bragg band gaps for surface guided waves in a phononic crystal of pillars. Phys. Rev. B Condens. Matter Mater. Phys. (2011)
6. M. Rupin, F. Lemoult, G. Lerosey, P. Roux, Experimental demonstration of ordered and disordered multiresonant metamaterials for Lamb waves. Phys. Rev. Lett. (2014)
7. A.A. Maradudin, *Structured Surfaces as Optical Metamaterials*. Dover Books on Engineering (2011)
8. E. Baravelli, M. Ruzzene, Internally resonating lattices for bandgap generation and low-frequency vibration control. J. Sound Vib. (2013)
9. M. Miniaci, A. Marzani, L. Testoni, L. De Marchi, Complete band gaps in a polyvinyl chloride (PVC) phononic plate with cross-like holes: numerical design and experimental verification. Ultrasonics (2015)
10. H. Lee, J.H. Oh, H.M. Seung, S.H. Cho, Y.Y. Kim, Extreme stiffness hyperbolic elastic metamaterial for total transmission subwavelength imaging. Sci. Rep. (2016)

11. K.H. Matlack, A. Bauhofer, S. Krödel, A. Palermo, C. Daraio, Composite 3D-printed metastructures for lowfrequency and broadband vibration absorption. Proc. Natl. Acad. Sci. USA (2016)
12. D. Tallarico, N.V. Movchan, A.B. Movchan, D.J. Colquitt, Tilted resonators in a triangular elastic lattice: chirality, Bloch waves and negative refraction. J. Mech. Phys. Solids (2017)
13. P.I. Galich, N.X. Fang, M.C. Boyce, S. Rudykh, Elastic wave propagation in finitely deformed layered materials. J. Mech. Phys. Solids (2017)
14. Y. Pennec, B. Djafari-Rouhani, H. Larabi, J.O. Vasseur, A.C. Hladky-Hennion, Low-frequency gaps in a phononic crystal constituted of cylindrical dots deposited on a thin homogeneous plate. Phys. Rev. B Condens. Matter Mater. Phys. (2008)
15. A. Colombi, P. Roux, S. Guenneau, P. Gueguen, R.V. Craster, Forests as a natural seismic metamaterial: Rayleigh wave bandgaps induced by local resonances. Sci. Rep. (2016)
16. A. Colombi, R.V. Craster, D. Colquitt, Y. Achaoui, S. Guenneau, P. Roux, M. Rupin, Elastic wave control beyond band-gaps: shaping the flow of waves in plates and half-spaces with subwavelength resonant rods. Front. Mech. Eng. (2017)
17. E.G. Williams, P. Roux, M. Rupin, W.A. Kuperman, Theory of multiresonant metamaterials for A0 Lamb waves. Phys. Rev. B Condens. Matter Mater. Phys. (2015)
18. A. Weinmann, L.D. Landau, E.M. Lifshitz, J.B. Sykes, J.S. Bell, Quantum mechanics (Nonrelativistic theory). Math. Gaz. (1959)
19. N.C. Perkins, C.D. Mote, Comments on curve veering in eigenvalue problems. J. Sound Vib. (1986)
20. V. Romero-García, R. Picó, A. Cebrecos, V.J. Sánchez-Morcillo, K. Staliunas, Enhancement of sound in chirped sonic crystals. Appl. Phys. Lett. (2013)
21. A. Cebrecos, R. Picó, V.J. Sánchez-Morcillo, K. Staliunas, V. Romero-García, L.M. Garcia-Raffi, Enhancement of sound by soft reflections in exponentially chirped crystals. AIP Adv. (2014)
22. A. Colombi, D. Colquitt, P. Roux, S. Guenneau, R.V. Craster, A seismic metamaterial: the resonant metawedge. Sci. Rep. (2016)
23. K.L. Tsakmakidis, A.D. Boardman, O. Hess, Trapped rainbow storage of light in metamaterials. Nature (2007)
24. J. Zhu, Y. Chen, X. Zhu, F.J. Garcia-Vidal, X. Yin, W. Zhang, X. Zhang, Acoustic rainbow trapping. Sci. Rep. (2013)
25. L. Rayleigh, On waves propagated along the plane surface of an elastic solid. Proc. Lond. Math. Soc. (1885)
26. J.H. Page, A. Sukhovich, S. Yang, M.L. Cowan, F. Van Der Biest, A. Tourin, M. Fink, Z. Liu, C.T. Chan, P. Sheng, Phononic crystals. Phys. Status Solidi (B) Basic Res. (2004)
27. A. Sukhovich, L. Jing, J.H. Page, Negative refraction and focusing of ultrasound in two-dimensional phononic crystals. Phys. Rev. B Condens. Matter Mater. Phys. (2008)
28. Z. Liu, X. Zhang, Y. Mao, Y.Y. Zhu, Z. Yang, C.T. Chan, P. Sheng, Locally resonant sonic materials. Science (2000)
29. Y. Achaoui, V. Laude, S. Benchabane, A. Khelif, Local resonances in phononic crystals and in random arrangements of pillars on a surface. J. Appl. Phys. (2013)
30. F. Lemoult, M. Fink, G. Lerosey, Acoustic resonators for far-field control of sound on a subwavelength scale. Phys. Rev. Lett. (2011)
31. A.E. Miroshnichenko, S. Flach, Y.S. Kivshar, Fano resonances in nanoscale structures. Rev. Mod. Phys. (2010)
32. N. Kaina, M. Fink, G. Lerosey, Composite media mixing Bragg and local resonances for highly attenuating and broad bandgaps. Sci. Rep. (2013)
33. R. Porter, D.V. Evans, Rayleigh-Bloch surface waves along periodic gratings and their connection with trapped modes in waveguides. J. Fluid Mech. (1999)
34. D.J. Colquitt, R.V. Craster, T. Antonakakis, S. Guenneau, Rayleigh-Bloch waves along elastic diffraction gratings. Proc. R. Soc. A: Math. Phys. Eng. Sci. (2015)
35. G.J. Chaplain, M.P. Makwana, R.V. Craster, Rayleigh–Bloch, topological edge and interface waves for structured elastic plates. Wave Motion (2019)

36. A. Colombi, V. Ageeva, R.J. Smith, A. Clare, R. Patel, M. Clark, D. Colquitt, P. Roux, S. Guenneau, R.V. Craster, Enhanced sensing and conversion of ultrasonic Rayleigh waves by elastic metasurfaces. Sci. Rep. (2017)
37. G.J. Chaplain, J.M. De Ponti, A. Colombi, R. Fuentes-Dominguez, P. Dryburg, D. Pieris, R.J. Smith, A. Clare, M. Clark, R.V. Craster, Tailored elastic surface to body wave Umklapp conversion. Nat. Commun. (2020)
38. N.W. Von Ashcroft, N.D. Mermin, *Solid State Physics, New York 1976, XXII, 826 Seiten* (Holt, Rinehart and Winston, 1976)
39. R. Peierls, Zur kinetischen Theorie der Wärmeleitung in Kristallen. Annalen der Physik (1929)
40. L. Brillouin, Wave propagation in periodic structures: electric filters and crystal lattices. Nature (1946)
41. C. Kittel, *Introduction to Solid State Physics*, 8th edn. (Wiley, New York, 2004)
42. L. Hoddeson, G. Baym, M. Eckert, The development of the quantum-mechanical electron theory of metals: 1928. Rev. Mod. Phys. (1987)

Chapter 5
Advanced Multiresonator Designs for Energy Harvesting

Abstract This chapter, starting from basic concepts on piezoelectric materials, explores piezo-augmented arrays of resonators able to dramatically increase the energy available for harvesting, and the operational bandwidth. Specifically, three designs are proposed, exploiting rainbow reflection, trapping, and topological edge modes.

5.1 An Introduction to Piezoelectricity

Piezoelectricity is the ability of a material to develop an electric charge in response to an applied mechanical stress (direct piezoelectric effect) and vice-versa (inverse piezoelectric effect). The term comes from the Greek words $\pi\iota\acute{\epsilon}\zeta\epsilon\iota\nu$, which means to squeeze or press and $\acute{\eta}\lambda\epsilon\kappa\tau\rho o\nu$, meaning amber, an ancient source of electric charge. It was firstly discovered by the french physicists Jacques and Pierre Curie in 1880, which demonstrated the direct piezoelectric effect in quartz and in other crystalline materials in the natural state.

The piezoelectric effect is due to the peculiar crystalline structure of such materials, with no inversion symmetry. Electrical dipoles within the piezoelectric material are responsible for the creation of a potential difference across the material, when the top and bottom layers are connected to electrodes. When the material is in the unstressed state, it is neutrally charged, since the positive and negative charges balance each other. Contrary, the application of a stress, changes the position of the charges, thus modifying the dipole moment (Fig. 5.1).

The most common piezoelectric materials are ceramics, and specifically Aluminium Nitride (AlN) and Lead Zirconate Titanate (PZT) due to their piezoelectric and manufacturability qualities. In order to describe the behaviour of piezoelectric materials in the setting of continuum mechanics, the electromechanical coupling enters in the constitutive laws [1]:

J. M. De Ponti, *Graded Elastic Metamaterials for Energy Harvesting*, PoliMI SpringerBriefs, https://doi.org/10.1007/978-3-030-69060-1_5

Fig. 5.1 Piezoelectric effect in quartz. When the material is in the unstressed state, positive and negative charge balance each other, while the application of a stress modify the dipole moment

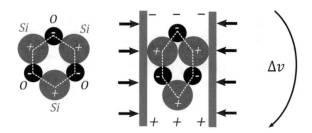

$$\begin{cases} T_{ij} = c^E_{ijkl} S_{kl} - e_{kij} E_k \\ D_i = e_{ikl} S_{kl} + \varepsilon^S_{ik} E_k \end{cases} \qquad (5.1)$$

where c^E_{ijkl} is a 4th order elastic stiffness symmetric tensor evaluated at constant electric field; e_{kij} is a 3rd order tensor of the piezoelectric stress constants, and ε^S_{ik} is a 2nd order tensor of the dielectric constants at constant strain. Equation (5.1) is the *e-form* of the piezoelectric constitutive equations, in which the strain and the electric field are used as coupling variables. Alternatively, it is possible to use the stress instead of the strain, obtaining the *d-form* of the piezoelectric constitutive equations:

$$\begin{cases} S_{ij} = s^E_{ijkl} T_{kl} + d_{kij} E_k \\ D_i = d_{ikl} T_{kl} + \varepsilon^T_{ik} E_k \end{cases} \qquad (5.2)$$

where s^E_{ijkl}, d_{kij} and ε^T_{ik} are the 4th order elastic symmetric compliance tensor at constant electric field, the 3rd order tensor of the piezoelectric strain constants and the 2nd order tensor of the dielectric constants at constant stress respectively.

The peculiarity of piezoelectricity is given by 3rd order tensors e_{kij} and d_{kij}, that couple mechanical and electrical quantities.

These coefficients refer to three main coupling mechanisms, as shown in Fig. 5.2, assuming that direction 3 is always the polarization one [2]. Specifically, the following modes can be identified:

- 33-mode: the application of an electric field along the polarization axis, stretches the piezoelectric element in the same direction (and vice-versa);
- 31-mode: the application of an electric field along the polarization axis, shrinks the piezoelectric element in the orthogonal plane (and vice-versa);
- shear-mode: the application of an electric field orthogonal to the polarization axis, deforms the piezoelectric element in shear (and vice-versa).

Only few coupling constant are non zero: e_{333} related to the 33-mode, $e_{311}=e_{322}$ to the 31-mode, and $e_{113}=e_{223}$ to the shear-mode. The constitutive equations are usually represented using the Voigt's notation, expressing 2nd order symmetric tensors with vectors, and 3rd and 4th order tensors with matrices. This is done by labeling the subscripts as: $11 \rightarrow 1, 22 \rightarrow 2, 33 \rightarrow 3, 13 \rightarrow 4, 23 \rightarrow 5, 12 \rightarrow 6$. In this way, the piezoelectric coupling matrix reduces to:

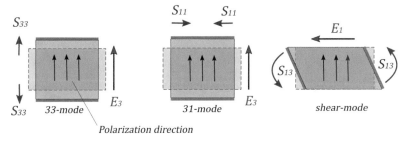

Fig. 5.2 Piezoelectric 33-mode, 31-mode and shear mode, assuming always 3 as the polarization direction

$$e = \begin{bmatrix} 0 & 0 & 0 & 0 & e_{15} & 0 \\ 0 & 0 & 0 & e_{24} & 0 & 0 \\ e_{31} & e_{32} & e_{33} & 0 & 0 & 0 \end{bmatrix} \tag{5.3}$$

while the symmetric elastic stiffness matrix to:

$$c^E = \begin{bmatrix} c_{11}^E & c_{12}^E & c_{13}^E & 0 & 0 & 0 \\ c_{12}^E & c_{22}^E & c_{23}^E & 0 & 0 & 0 \\ c_{13}^E & c_{23}^E & c_{33}^E & 0 & 0 & 0 \\ 0 & 0 & 0 & c_{44}^E & 0 & 0 \\ 0 & 0 & 0 & 0 & c_{55}^E & 0 \\ 0 & 0 & 0 & 0 & 0 & c_{66}^E \end{bmatrix} \tag{5.4}$$

5.2 Graded Metasurface for Energy Harvesting

A graded resonator array, as the one proposed in the previous chapter (Fig. 4.8), is able to concentrate energy using the rainbow effect. Because such systems already contain a collection of resonators, the inclusion of vibrational energy harvesters is straightforward leading to truly multifunctional metasurfaces combining vibration insulation with harvesting, as explained in [3].

Concentrating energy at a known spatial position, as the graded array does, is only part of the requirement for harvesting: it is also necessary to design an arrangement for the piezoelectric patches that takes full advantage of the displacements induced within the array and upon the rods. To understand the mechanisms involved we will take a single piezo-augmented resonator and place it within the graded array, as shown in Fig. 5.3a.

Recalling that the dominant mode of interest is the longitudinal one, the harvester we use is a rod with four cantilever beams arranged in a cross-like shape placed upon the top. Each beam embeds a piezoelectric patch and their motion in harvesting mode is shown in Fig. 5.3b. Both the supporting beam and the rods are made of aluminium

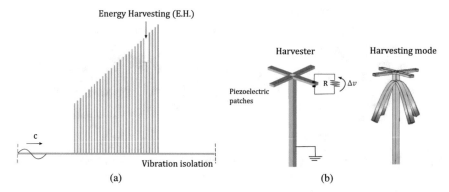

Fig. 5.3 **a** Schematic of the graded metasurface able to simultaneously provide vibration isolation and energy harvesting. **b** Detailed view of the harvester, with the piezoelectric patches, electric circuit, and corresponding harvesting mode based on axial elongation of the rod and flexure of the cantilevers

($E_a = 70$ GPa, $v_a = 0.33$ and $\rho_a = 2710$ kg/m^3), while the piezoelectric patches are made of PZT-5H ($E_p = 61$ GPa, $v_p = 0.31$ and $\rho_p = 7800$ kg/m^3). The harvester design (Fig. 5.3b) has strong dynamic coupling between the rod and the cantilever beams as we carefully design it to work in the double amplification regime coupling the axial fundamental mode with the flexural one of the beam, i.e. the lengths of the cantilever beams are not arbitrary. The design values are: $A_r = 25$ mm^2 (square cross section), $A_b = 10$ mm^2 (same width as the rod, i.e. 5 mm), $l_r = 460$ mm, $l_b = 25$ mm and $h_p = 0.3$ mm (the subscript r denotes the properties of the rod, b of the beams and p of the patches). This harvester design provides a multimodal response of the system, always exploiting the fundamental flexural mode of the beams.

The fundamental mode of the cantilever beam is that most suited for energy harvesting as it does not result in charge cancellation and has the highest power production; with higher mode numbers this would decay by around two orders of magnitude [1, 4]. In addition, this configuration provides high amplification at low frequency (relevant for ambient vibrations) due to both the slenderness of the rod and the added mass from the four cantilever beams. The harvester is designed using a simple, yet highly effective, mass-spring model. We use a 1D spring-mass model, inset in Fig. 5.4, and adopt Timoshenko beam theory.

Defining with $w_r(z, t)$ and $w_b(x, t)$ the vertical displacement of the rod and the beam respectively, the effective mass is obtained writing the kinetic energy of the system for both the axial and flexural mode as:

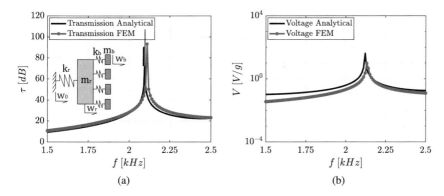

Fig. 5.4 Comparison between the analytical (model in inset) and numerical transmission spectrum in short circuit (**a**) and voltage in open circuit (**b**)

$$T_r(t) = \frac{1}{2} M_b \left[\frac{\partial w_r(z,t)}{\partial t} |_{z=l_r} \right]^2 + \frac{1}{2} \int_{z=0}^{z=l_r} \rho_a A_r \left[\frac{\partial w_r(z,t)}{\partial t} \right]^2 dz = \frac{1}{2} \left[M_b + \frac{1}{3} \rho_a A_r l_r \right] \dot{w}_r^2(t),$$
(5.5)

$$T_b(t) = \frac{1}{2} \int_{x=0}^{x=l_b} (\rho_a A_b + \rho_p A_p) \left[\frac{\partial w_b(x,t)}{\partial t} \right]^2 dx = \frac{1}{2} \left[\frac{33}{140} (\rho_a A_b + \rho_p A_p) l_b \right] \dot{w}_b^2(t), \quad (5.6)$$

while elastic and electric lumped coefficients directly follow from the internal energy definition:

$$U_r(t) = \frac{1}{2} \int_{z=0}^{z=l_r} \int_A (T_{zz} S_{zz}) \, dA \, dz = \frac{1}{2} k_r w_r(t)^2,$$
(5.7)

$$U_b(t) = \frac{1}{2} \int_{x=0}^{x=l_b} \int_A (T_{xx} S_{xx} + T_{xz} S_{xz} - D_{zz} E_{zz}) \, dA \, dx = \frac{1}{2} k_b w_b^2(t) - \theta w_b(t) v_b(t) - \frac{1}{2} C_p v_b^2(t),$$
(5.8)

with a separation of variables adopted as (assumed mode method): $w_r(z,t) = w_r(t) z/l_r$ and $w_b(x,t) = \frac{1}{2}[3(x/l_b)^2 - (x/l_b)^3] w_b(t)$, and being $M_b = 4(\rho_a A_b + \rho_p A_p) l_b$ the total mass of the four beams, $\underline{T}$ and $\underline{S}$ the stress and strain fields, θ the electromechanical coupling coefficient and C_p the internal piezoelectric capacitance. Lumped coefficients are then defined as: $k_r = E_a A_r / l_r$, $k_b = 1/(\frac{l_b^3}{3 E_a I_b} + \frac{l_b}{G_a A_b^s})$, $\theta = -\frac{3 e_{31} a_p}{2 h_p l_p} (h_p^2 + 2 b_b h_p - 2 y_n h_p)$, $C_p = \bar{\varepsilon}_{33} a_p l_p / h_p$, $m_r = 1/3 \rho_a A_r l_r + M_b$, $m_b = (33/140)(\rho_a A_b + \rho_p A_p) l_b$ and being $G_a = E_a/2(1 + \nu_a)$ the aluminium shear modulus, a_p the patch width, b_b the beam thickness, y_n the neutral axis of the composite beam cross section, e_{31} the considered piezoelectric coefficient and $\bar{\varepsilon}_{33}$ the constant-stress dielectric constant. The dynamics, of the electromechanical problem, are now succinctly described by three linear coupled ordinary differential equations:

$$\begin{cases} m_r \ddot{w}_r + k_r(w_r - w_0) + 4k_b(w_r - w_b) = 0, \\ m_b \ddot{w}_b + k_b(w_b - w_r) - \theta v_b = 0, \\ C_p \dot{v}_b + v_b/R + \theta \dot{w}_b = 0, \end{cases} \tag{5.9}$$

where R is the electrical resistance. Fourier transforming Eq. (5.9) gives the transfer function as:

$$\tilde{T}(\omega) = \frac{k_r k_b}{(-m_r \omega^2 + k_r + 4k_b)(-m_b \omega^2 + k_b + \frac{i\omega\theta^2}{i\omega C_p + 1/R}) - 4k_b^2}, \tag{5.10}$$

from which the voltage and power follows directly as:

$$\tilde{V}(\omega) = -\frac{i\omega\theta}{i\omega C_p + 1/R} \tilde{T}(\omega) w_0, \qquad \tilde{P}(\omega) = \frac{1}{R}\left[\frac{i\omega\theta}{i\omega C_p + 1/R} \tilde{T}(\omega) w_0\right]^2$$

with w_0 the imposed harmonic displacement amplitude. For brevity we do not incorporate damping, but it can be easily included in Eq. (5.9) by introducing a term linearly dependent on the velocity $\dot{w}$. The main contributor to damping in this system is provided by the piezoelectric material since the quality factor of aluminium is very high. However we neglected damping here since the thickness of the patch is very small with respect to that of the beam.

This simplified model provides surprisingly accurate predictions in the frequency range of interest (see Fig. 5.4a, b), with just a 0.7% of error in the natural frequency prediction.

The product RC_p defines the time constant of the circuit τ_{RC}, providing measure of the time required to charge the capacitor through the resistor. The time constant is related to the cutoff circular frequency which is an alternative parameter of the RC circuit:

$$\omega_{RC} = \frac{1}{\tau_{RC}} = \frac{1}{RC_p}. \tag{5.11}$$

The values of R maximising the electric power at each excitation frequency are obtained by imposing its stationarity with respect to R (the dashed white line in Fig. 5.5a):

$$R_{opt.}(\omega) = \frac{-k_r k_b + k_r m_b \omega^2 + k_b m_r \omega^2 + 4k_b m_b \omega^2 - m_r m_b \omega^4}{k_r \omega \theta^2 + 4k_b \omega \theta^2 - m_r \omega^3 \theta^2 - C_p k_r m_b \omega^3 - C_p k_b m_r \omega^3 - 4C_p k_b m_b \omega^3 + C_p m_r m_b \omega^5 + C_p k_r k_b \omega}. \tag{5.12}$$

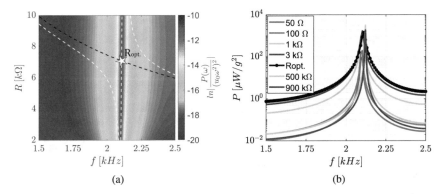

(a)   (b)

Fig. 5.5 Power output versus electrical resistance and frequency (**a**). Dashed white line shows the optimal loading at each excitation frequency. Dashed black line the cutoff frequency for different values of resistance. **b** show the corresponding electrical resistance for optimal power (normalized with respect to the gravitational acceleration g^2) (blue curve) compared with other resistance values

The optimal electrical resistance is then obtained from the intersection of Eqs. (5.11) and (5.12) (the intersection of black and white dashed lines in Fig. 5.5a—white star), obtaining $R_{opt} = 7.08\,k\Omega$.

Now that we have obtained a harvester optimised for the longitudinal rod resonance, it can be introduced into the metasurface array to assess whether its performance is increased when the rainbow effect occurs. We present numerical and experimental results of the graded array on an elastic beam. The array is attached to an aluminium beam 30 mm wide and of thickness of 10 mm which is sufficiently stiff to avoid anomalous resonances in the rod. The array is composed of 30 rods, each with square cross section of area 25 mm^2, a linear height gradient ranging from 250 to 650 mm and a constant spacing between the rods of 15 mm; this results in a ~43° slope angle. Equation (4.46) suggests that the metasurface will have the longitudinal fundamental mode in the range 2–5 kHz and rod number 26 (with the rod numbering in the array having 1 the shortest resonator and 30 the longest), with length 460 mm, is the harvester previously designed.

The graded metasurface is illustrated in Fig. 5.6, where the absorbing boundaries are defined using a symmetric ABH, obtained by gradually varying the beam thickness and through the addition of a dissipating material [6]. The energy harvesting performance of the device is quantified connecting each piezoelectric patch to an independent resistive load and measuring the electric power (see Fig. 5.6).

The experimental setup is reported in Fig. 5.7, where the array of resonators is clamped on the host beam through a set of screws. The applied torque slightly modifies the axial resonance frequency of the rods which is reduced of approximately 100 Hz with respect to the numerical prediction. Therefore, the device is tested at 2.05 kHz instead of 2.15 kHz, in order to properly excite the resonance frequency of the harvester.

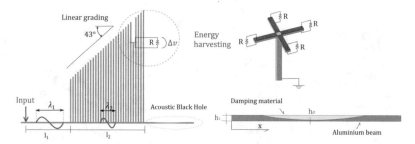

Fig. 5.6 Schematic of the system. The rainbow device is made of an array of resonant rods covering a length $l_2 = 435$ mm, which are clamped on a host beam. The input region is represented in red and it is located at a distance $l_1 = 300$ mm from the array. The output region, corresponding to a rod equipped with a harvester, is represented in blue. In the right part of the figure, a detailed view of the harvester and the ABH is shown

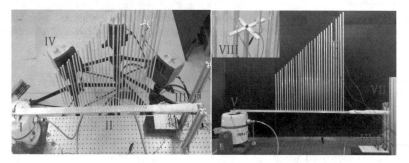

Fig. 5.7 Experimental setup. The excitation is provided through an electrodynamic shaker (I). The rainbow metasurface is mounted on an otherwise plain beam (II) supplemented by an acoustic black hole (III), to prevent spurious reflections, and a 3D laser vibrometer (IV) is used to measure the wavefield on the bottom surface of the beam. The input acceleration is measured through an accelerometer located alongside the shaker (V) while the output power is measured by connecting the piezo-electrodes to a passive resistive load (VI). The system is suspended through elastic cables located alongside the acoustic black hole (VII). A zoomed-in view of the harvester is shown in (VIII)

The electro-mechanical system is made of a beam augmented with 30 resonators, which are mounted on the beam through a set of screws, emulating rigid connections. The beam is also mechanically joined to a $LDSv406$ electrodynamic shaker at the left boundary, to provide an out-of-plane input excitation able to activate the A_0 Lamb mode; the beam is placed perpendicularly to the excitation. An accelerometer, $PCB352C33$, is bonded to the beam to measure the acceleration level provided by the shaker and data is acquired using a National Instruments (NI) input module with internal signal conditioning. The opposite end is attached to an external frame via a set of ropes, to avoid undesired deformations of the beam; the presence of the ropes is assumed not to affect the system dynamics. In addition, an ABH, is manufactured at the right-hand boundary to minimize any reflection of waves and to emulate absorbing boundaries. To emulate absorbing boundaries we consider an

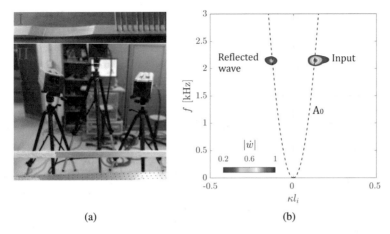

(a)   (b)

Fig. 5.8 **a** Experimental setup for the characterization of the ABH. The beam is clamped on the left side and the wavefield is measured on the bottom surface using a 3D scanner laser Doppler vibrometer (SLDV). The acoustic black hole is manufactured by gradually reducing the beam thickness of an otherwise constant-thickness beam and through the addition of a highly dissipative material in correspondence of the reduced cross section (I). **b** Experimental dispersion curve illustrating the input and reflected A_0 Lamb modes

ABH similar to [7–9], and illustrated in Fig. 5.8a-I. Specifically, the beam cross-section is machined to achieve a variable profile for the thickness characterised by the following expression:

$$h(x) = \varepsilon x^m + h_0, \tag{5.13}$$

where $m = 2.2$, $h_0 = 2$ mm and $\varepsilon = 1.6$. The addition of a highly dissipative material ($Blu - Tack$) to the variable section provides localised damping during the wave propagation, that allows to avoid undesired reflections at the beam's free edge, therefore emulating absorbing boundary conditions. We experimentally verified the behavior of the acoustic black hole through the setup illustrated in Fig. 5.8a-II, in which only the host beam is mounted on the optical table. The aforementioned input forcing signal ($f = 2.05$ kHz, $\Delta f = 0.07$ kHz) is imposed at the boundary opposite to the black hole. The experimental dispersion relation $|\dot{w}(\kappa, f)|$ is computed through the *Fourier Transform* (FT) of the velocity field and compared to the analytical dispersion for the A_0 Lamb mode (Eq. (3.16)) in Fig. 5.8b. As expected, the energy content is mainly located on the positive wavenumber domain, whereas only a small amount of energy (almost 50% of reflected component) is present in the left side.

Once verified the ABH capability to reduce spurious edge reflections, we investigate the metasurface influence on the wave propagation and energy harvesting. The 26th resonator, that is placed at the coordinate at which the wave speed at 2.05 kHz is minimum, is equipped with a harvester, able to convert the mechanical vibrations into electric energy. The piezoelectric harvester is made of four cantilever beams arranged in a cross-like shape (see Fig. 5.6). Each of them is made of an aluminum substrate of 25 mm length, 5 mm width and 2 mm thickness, endowed with a piezoelectric PZT-5H patch ($E = 61$ GPa, $v_p = 0.31$, $\rho_p = 7800$ kg/m^3, dielectric constant $\varepsilon_{33}^T/\varepsilon_0 = 3500$ and piezoelectric coefficient $\varepsilon_{31} = -9.2$ C/m^2) of the same in plane dimensions, with a cross-section thickness of $t = 0.3$ mm and bonded through a conductive epoxy $CW2400$.

A Polytec 3D scanner laser Doppler vibrometer (SLDV) is used to measure the velocity field along the main dimension of the supporting beam, separating the out-of-plane component. At the same time, the voltage across the harvester's electrodes is acquired through the external Polytec acquisition system. Finally, the excitation signal (provided through a $KEYSIGHT$ 33500B waveform generator), synchronously starts with the measurement system and consists in a input function $V(t) = V_0 w(n) sin(2\pi f_c t)$ with amplitude $V_0 = 5$ V, $w(n)$ is a Hann window, central frequency $f_c = 2.05$ kHz and time duration 30 ms; this results in a spectral content of width $\Delta f = 0.14$ kHz. Two different testing conditions are considered: (a) only one resonator, equipped with the harvester, is placed on the beam; (b) all the resonators are present, whereby the metasurface enables the wave speed control. This enables us to quantify the difference between a lone single harvester and the harvester embedded within the metasurface. The resulting velocity field, for the single rod, is illustrated in Fig. 5.9a-I; since the beam is characterized by constant material properties along its main dimension, the associated wavelength is invariant in space, as is clearly visible in the velocity profile in Fig. 5.9a-II, corresponding to the time instant illustrated with the horizontal dashed line. In addition, at a generic point belonging to the beam, the imposed wave drops to zero after approximately 25 ms, which is highlighted in the vertical cut of the velocity field shown in Fig. 5.9a-III.

The corresponding spectrogram is shown in Fig. 5.9b; it is computed through a 2D *Fourier Transform* (FT) of the velocity field, properly windowed with a moving Gaussian function along x, which results in the function $\hat{v}(\kappa, x, f)$. The dependence upon frequency is eliminated by taking the RMS value in time, which allows to define the spectrogram as the amplitude $|\hat{v}(\kappa, x)|$, displayed with colored contours in Fig. 5.9b, confirming that wave propagation occurs without wavenumber transformation [10]. Turning to the harvester embedded within the metasurface, the effect of the metasurface is visible in Fig. 5.9c-I and in the corresponding horizontal (Fig. 5.9c-II) and vertical (Fig. 5.9c-III) cuts. Interestingly, the associated wavelength varies along the x-direction, whereby its variation is accompanied by amplitude decrease, as expected from the arguments presented earlier. We also observe that the temporal response is delayed and that the energy remains in the system for a longer time compared to the single harvester, which further enhances the metasurface effect. This effect is created by the partial wave scattering and wave confinement occurring inside the array, and this is further confirmed by comparison between the expected wavenumber transformation (obtained with the numerical model and represented with black dashed lines) and the experimental spectrogram in Fig. 5.9d.

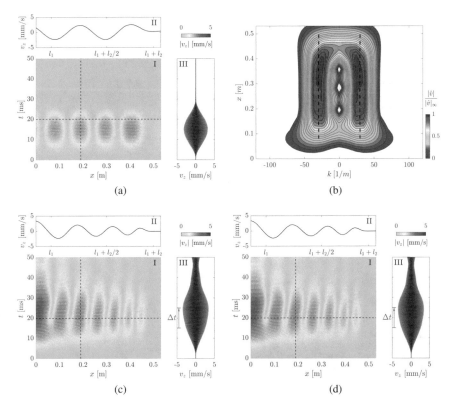

Fig. 5.9 Experimental velocity field for a beam with **a** one resonating rod and **c** the metasurface. The horizontal dashed lines are representative of the time instant $t = 20$ ms, corresponding to the wave profile displayed on the top of the figure (II); at this time instant, the typical pattern of nodes and antinodes characterising a standing wave is visible. The vertical dashed line is representative of the temporal wave profile measured at $x = 0.2$ m and illustrated in the right side of the figure (III). Corresponding spectrograms for **b** single resonator and **d** the metasurface, with superimposed numerical dispersion for constant frequency and different resonator height (dashed black line). While for the single harvester the wavenumber is constant in space, a relevant wavenumber transformation can be observed when the metasurface is mounted on the beam [6]

The experimental results are corroborated numerically (see Fig. 5.10) using Abaqus, performing a time domain implicit analysis with a constant time increment $dt = 5$ μs. The system is forced using the imposed acceleration that is experimentally measured with the accelerometer placed in correspondence of the shaker, while the black hole is modeled numerically by varying the beam profile and adding a soft material with Rayleigh damping $\alpha = 5000$ rad/s. In addition, we add in the harvester model a Rayleigh damping source, with $\alpha = 19.3$ rad/s and $\beta = 2.93 \times 10^{-6}$ s/rad, on the basis of conventional modal damping ratios [1] for aluminum beams with piezoelectric patches.

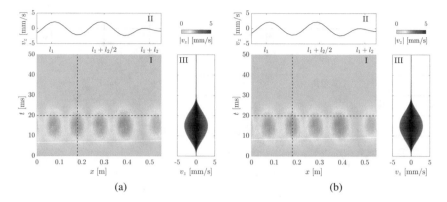

(a)   (b)

Fig. 5.10 Numerical velocity field computed for **a** beam with one resonator; **b** beam with graded metasurface. The horizontal and vertical dashed lines in (I) correspond to the reference coordinate and time instant employed to illustrate the wave behavior in space (II) and time (III), as shown on top and alongside the velocity field [6]

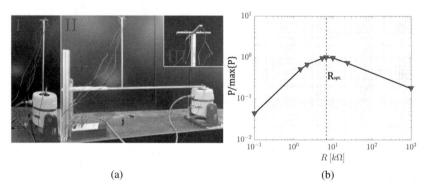

(a)   (b)

Fig. 5.11 **a** Experimental setup adopted for the evaluation of the optimal electric load and associated electric circuit with parallel connection (I). The same load and electric connections are employed in the experimental tests of the beam with attached resonating rod (II) and beam with metasurface. Zoom-in view of the harvester (III). **b** Representation of the peak power for different resistive loads. The optimum values is approximately 6.8 kΩ

To optimize the power extracted during the tests, a passive resistor is selected and connected to each piezo-patch based on the experimental characterization of the dynamical response. The analysis is performed through the experimental setup displayed in Fig. 5.11a-I, where the resonator is placed vertically on the shaker. The system is excited using a sweep sine in the neighborhood of the resonance frequency of the cantilever and the rod (which are carefully designed for operating at the same frequency). The voltage drop across the piezoelectric electrodes is measured upon varying the resistance values, whereby the power achieved at 2.05 kHz is displayed in Fig. 5.11b normalized by the its maximum.

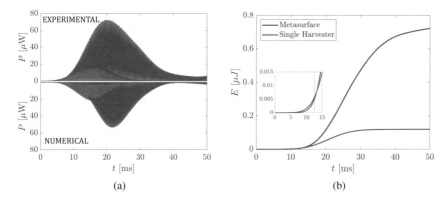

Fig. 5.12 **a** Numerical and experimental electric power produced by the harvester with (blue) and without (red) the metasurface. **b** Experimental cumulated electric energy in time

The analysis reveals an optimal resistance of 6.8 kΩ, which is in agreement with the lumped model computations of the previous section.

Finally, we quantify the energy harvesting capabilities of the system by comparing the experimental power output for the host beam with one resonator (red line) and the power achieved through the metasurface effect (blue line), as shown in Fig. 5.12a (top), for the same excitation level. It is demonstrated that, thanks to the careful design of the array of rods, the wave propagation is slowed down. As a result, a greater interaction time between the wave and the harvester and a wave localization responsible for a field amplification on the rod endowed with the harvester is guaranteed. This process reflects into a greater and delayed power output. In addition to these experimental results, a comparison with the power calculated through a numerical simulation in Abaqus is shown in Fig. 5.12a (bottom), imposing the experimental acceleration time history. The output voltage is displayed in the bottom part of Fig. 5.12, for the case of one resonator (red) and of a full metasurface (blue). The agreement between the experimental and the numerical data confirms that the rainbow effect is the observed mechanism and that the system is operating as we predict. The slightly different behaviour between the experimental and numerical EH results could be attributed to several aspects, mainly connected to the effect of imperfections (rod heights, tightening of the screws) and to the experimental spurious reflections of the acoustic black hole that increase the energy confinement in the host beam. As a matter of fact, the experimental/numerical agreement is more than satisfactory for the case of a single resonator, demonstrating the correctness of the model. For the case of the array, in which the number and the interplay of imperfections is by far larger than before, some differences arise.

The accumulated energy for the two scenarios, the lone harvester and the harvester embedded in the metasurface, are then compared in Fig. 5.12b in the range 0–50 ms, which corresponds to a total energy of 0.12 and 0.72 μJ, demonstrating a strong increase of the harvesting capabilities. The advantage of the system is evident only after a certain amount of time, and benefits from a sufficiently continuous, or long duration, source of excitation (see inset in Fig. 5.12b); if the excitation is discontinuous or very short, the metasurface is less efficient. On the other hand, the broadband behaviour of the device can be enhanced by placing more harvesters as in [3].

5.3   From Rainbow Reflection to Rainbow Trapping

The rainbow effect, i.e. the spatial signal separation depending on frequency, can be obtained by exploiting zero group velocity modes at different frequencies along space. This can be done, like for the graded metasurface, by placing resonators of different height along an elastic substate. However, it is important to notice that the zero group velocity modes used to infer the rainbow phenomenon in discrete graded systems, can arise through a number of avenues [11]. Figure 5.13 shows the typical dispersion curves for three model systems were zero group velocity modes exist.

Figure 5.13a shows a flat dispersion branch, typical of slow sound or slow light in periodic systems composed of resonant elements, where over a large region in wavenumber space the group velocity is very low. This was the case of the graded metasurface, where the group velocity can be greatly reduced compared to a wave in a free beam. Figure 5.13b shows the dispersion for a (potentially non-resonant) periodic array, where a zero group velocity mode is achieved purely due to the Bragg condition being met at the edge of the first BZ; in a perfectly periodic medium standing waves form at this frequency due to reflections. These reflections prove key to delineating between true rainbow trapping and what we will term 'rainbow reflection' phenomena in accord with [11]. In Fig. 5.13c we show the dispersion curves for a symmetry broken array, where an accidental degeneracy is lifted, resulting in zero group velocity modes within the first BZ. We will focus on geometries capable of supporting such modes, which offer larger trapping potential due to the lack of coupling with reflected modes.

Whilst the positions of the zero group velocity modes in reciprocal space may seem a technicality, we show these nuances and their effect on the resulting wave phenomena have strong influences on energy harvesting applications for such arrays. Figure 5.13d–f shows some periodic configurations with dispersion curves according to Fig. 5.13a–c respectively. Figure 5.13d shows a resonant structure with the characteristic curve in Fig. 5.13a. These C-shaped resonators are used in water wave systems [12] and have analogues to split-ring resonators in electromagnetism, or mass-spring models analogous to rods atop an elastic halfspace [13]. Figure 5.13e, related to dispersion in Fig. 5.13b, shows a conventional periodic device called photonic/phononic crystals, depending on the regime being investigated. In acoustics this would be an array with acoustically rigid inclusions [14], or materials of varying dielectric constants in electromagnetism [15]. Finally, Fig. 5.13f, shows a symmetry broken array where accidental degeneracies are lifted resulting in zero group velocity modes inside the Brillouin zone, as in Fig. 5.13c. Such arrays are normally considered in identifying protected edge states within topological insulators [16, 17].

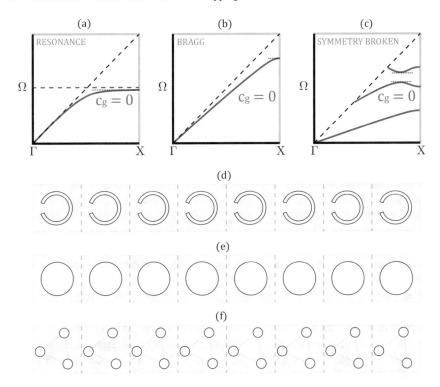

Fig. 5.13 Dispersion curves for differing periodic systems, with zero group velocity modes shown with dotted black lines. **a** Resonant system with zero group velocity mode induced by resonance (dashed blue line). Figure **d** shows a resonant structure with such a characteristic curve. **b** A (potentially non-resonant) system where zero group velocity modes are induced by the Bragg condition only, i.e. due to the periodicity. Figure **e** shows a conventional periodic device with such a dispersion branch, so called photonic/phononic crystals, depending on the regime being investigated. **c** Symmetry broken system, where accidental degeneracies are lifted resulting in zero group velocity modes inside the Brillouin zone. Figure **f** shows a symmetry broken array with such a characteristic dispersion relation. These structures have been realised in all imaginable wave systems, but here we focus resonant rods atop an elastic beam [11]

To illustrate the importance of the difference between these two effects, we propose a piezo-augmented array of resonators for harvesting electric energy from elastic substrates, and compare its functionalities with ungraded arrays and graded arrays capable of employing rainbow reflection and rainbow trapping. In order to differentiate between the two effects, the symmetry of the inclusions within the unit cell is of paramount importance. We emphasise that rainbow reflection can be achieved with any inclusion geometry, since leverages only the Bragg condition by virtue of the periodicity. For true rainbow trapping effects, trapping must be located at wavevectors within the first BZ, and hence rely on the decoupling of orthogonal eigensolutions, or symmetry breaking of the array geometry [16, 19–21] so to lift accidental degeneracies within the first BZ. Central to the design of rainbow trapping and reflecting arrays, is to adiabatically grade the array with respect to some set of parameters, thereby altering the local dispersion curves of the structure. The choice of the grading parameter can be any combination of the inclusion geometry, mass values, unit cell or loading element (for KL plates these can be pins, masses or resonators [22]). By altering such parameters, the dispersion curves of subsequent unit strips are pushed up or down in frequency, or indeed completely change their shape. For a given frequency, there will then be regions were a wave is either supported or prohibited from propagating; local band gaps are then reached at different spatial positions for different frequency components. At this position in a conventional perfectly periodic system standing waves form through subsequent Bragg reflections due to the periodicity. In graded systems, this has been labelled as rainbow trapping, since there is no forward propagating mode to couple into beyond this position, due to the encountered band gap. However, particularly in systems where no resonance effects are encountered [18], this mode is quickly reflected and couples to a counter propagating wave which travels along the array in the opposite direction. Indeed the reflected wave after this 'trapping' can be used for applications other than energy harvesting, such as flat lenses by passive self-phased effects [18].

The misnomer of rainbow trapping has been attributed to this effect in almost all graded systems where locally periodic dispersion curves are analysed, as the wave is seen to be prohibited from propagating further along the array. Due to the reduced speed of the wave, it can appear to stay localised for a considerable length of time [23], however unlike a truly trapped wave, this wave will ultimately reflect due to the position where the group velocity vanished.

The position of the reflection is frequency dependent through the grading parameter, and as such, we term this effect as 'rainbow reflection'. To distinguish between this reflection phenomenon and desired true rainbow trapping, we now utilise symmetry broken arrays. Degenerate eigensolutions of the dispersion relation, or accidental degeneracies (band crossings) within the Brillouin zone correspond to orthogonal modes and can exist if there is reflectional symmetry of the inclusion geometry about the array axis [21]. These more complex geometries are not normally analysed in graded systems, but the extension to such geometries is straightforward. Upon the breaking of the reflectional symmetry of the array, the degeneracy is lifted giving rise to solutions which are neither symmetric (even) or antisymmetric (odd) with respect to the array axis. Here the grading parameter is defined by the rotation through an angle $\Delta\theta = 2\pi/N$, where N is the number of unit cells within the graded region. In this way the rate of grading can be controlled, as well as how quickly the dispersion curves change.

We compare two graded line arrays composed of clusters of aluminium rods ($E = 70$ GPa, $\nu = 0.33$ and $\rho = 2710$ kg/m^3) atop an elastic beam, each with different gradings. The first exhibits conventional rainbow reflection; inspired by metawedge [24, 25] structures we design a one dimensional array of rods, with a single rod per unit cell, each increasing in height in subsequent cells (Fig. 5.14b). No symmetry induced accidental degeneracies arise in this setting, and as such to reach a zero group velocity mode, we must utilise the grading at the band edge by virtue of the Bragg condition. The second grading, incorporates the lifting of accidental degeneracies

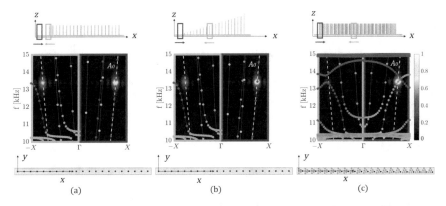

Fig. 5.14 Dispersion spectra for array of constant height (**a**) a metawedge (**b**) and rotated cell array (**c**), corresponding to total scattering, Rainbow reflection and Rainbow trapping at time 10 ms. Right panel shows spectra for the input wave, at position highlighted by purple rectangle in array schematics above, whilst the left panel shows the reflected wave spectra due to the band gap opening, at cell positions matching green boxes. For **a, b** the band gap opens via Bragg scattering, whilst in **c** due to the lifting of the accidental degeneracy. Overlaid on the spectra are the dispersion curves for highlighted cells; scatter points colours represent the wave polarization (purple corresponding to vertical motion, i.e. axial elongation). The arrays are excited through an A_0 Lamb wave at 13.5 kHz. Below each plot is an aerial view of the array [11]

through breaking inversion symmetry by smoothly rotating a triangular array of rods within a unit cell from $0°$ (i.e. symmetric about array axis) to $30°$. Both arrays are composed of 21 cells with 30 mm size and 10 mm thickness.

To quantify the reflection and trapping in the graded arrays, we analyse the dispersion spectra and compare both arrays with an ungraded periodic array of rods with equal number of unit cells. The ungraded array is composed of resonators of 80 mm height and 6 mm of diameter, similarly to the array with rotated cells (see Fig. 5.14a, c). The metawedge (Fig. 5.14b) is obtained by a linearly grading defined by a $16°$ slope angle, ranging in height from 2 to 175 mm. Comparison is performed through a time domain simulation in Abaqus, exciting the line array for 30 ms with an antisymmetric (A0) Lamb wave with a central frequency of 13.5 kHz, corresponding to the band gap opening for rods of height 80 mm and diameter 6 mm. Absorbing boundaries are imposed at the beam edges using the ALID method [5]. Dispersion curves are computed in Abaqus with a user defined code capable of imposing Bloch–Floquet boundary conditions. The input and reflected waves are obtained by applying a spatiotemporal Fourier transform on the wavefield before the array. The reflected wave, as a percentage of the incident radiation, at 10 ms (Fig. 5.14) is 71% for the ungraded array (a), 57% for the rainbow reflective device (b) and 21% for the symmetry broken rainbow trapping configuration (c). There is then a clear difference in mechanism for the slowing down and reflecting/trapping for the respective arrays; the energy is stored for longer in the trapping device. This difference is emphasised when increasing the observation temporal window from 10 to 30 ms (total input duration), we see the portion of reflected wave increases, reaching (Fig. 5.15), 91%

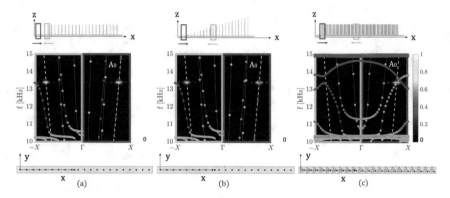

Fig. 5.15 Similar to Fig. 5.14, but at time of 30 ms (total input duration) [11]

(a), 83% (b) and 21% (c) of the input signal for the scattering, reflection and trapping cases respectively. Considering reference to the case with equal rods (no grading), the metawedge reduces the reflections to almost 20% at 10 ms and 9% at 30 ms, while the rotating cell grading of 70% at 10 ms and 77% at 30 ms. The reduction of reflection in the array with rotated cells is due to the band gap opening in a position far from the edge of the first BZ; this provides a longer interaction of the wave with the resonators, since the coupling to a reversed wave is less than in the case where the grading introduces reflection from the band edge. As is common when dealing with low loss, high Q-factor, aluminium structures, damping effects are neglected [13] throughout these simulations. Realistically we expect dissipation from the position of 'trapping' in all cases; the key difference for trapping devices, opposed to those which purely reflect, is that the time scale over which these effects dampen the slowed wave is much larger. As such trapping devices have a larger 'interaction time' with the device at the trapping position than their reflective counterparts.

In order to quantify the degree of trapping and the piezoelectric energy harvesting enhancement, we consider a piezoelectric disk, of 6 mm diameter and 2 mm thickness, at the base of each resonator for each cases of the ungraded array and the rainbow reflective metawedge configuration. To ensure a fair comparison between the symmetry broken triangular configuration, only one piezoelectric disk is considered per cell (highlighted and in aerial views in Figs. 5.14 and 5.15). This ensures for the three cases, the amount of piezoelectric material is exactly the same. As shown by the dispersion bands in Figs. 5.14 and 5.15, the unit cell for cases (a), (b) and (c) have been properly designed to obtain axial elongation of the rod with the piezo disk at the base at the frequency corresponding to the band gaps opening (13.5 kHz). Thus, we have a dominant d_{33} component of the piezoelectric tensor. The piezoelectric material is PZT-5H with piezoelectric coefficients $d_{31} = -274$pm V^{-1}, $d_{33} = 593$pm V^{-1}, $d_{15} = 741$pm V^{-1} and constant-stress dielectric constants $\varepsilon_{11}^T/\varepsilon_0 = 3130$ and $\varepsilon_{33}^T/\varepsilon_0 = 3400$ with $\varepsilon_0 = 8.854$ pF m^{-}1 the free space permittivity [1].

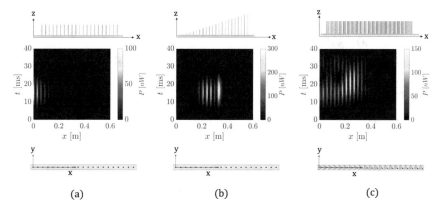

Fig. 5.16 Spatiotemporal power for the ungraded array (**a**) the metawedge (**b**) and the rotated cell array (**c**), corresponding to total scattering, rainbow reflection and rainbow trapping (input stops at 30 ms). The arrays are excited through an A_0 Lamb mode at 13.5 kHz. Each piezo disk is connected to a resistive load $R = 100$ kΩ [11]

In order to estimate the electric power transduced by the piezo-augmented arrays, each piezoelectric disk is attached to a resistive load $R = 100$ kΩ. Piezo disks are electrically independent (no series or parallel connections) in order to avoid possible charge cancellation due to out of phase responses. This is numerically modeled using Abaqus complemented with a customised Fortran subroutine as in [3]. Computing the electrical power, we see the maximum local value is obtained by the rainbow *reflection* array (see Fig. 5.16b), obtaining values approximately up to 300 nW. However, inspecting the time duration of the electric power production reveals that rainbow *trapping* has the longest period of power output (see Fig. 5.16c). The scattering array, i.e. the ungraded case, has the worst performance in both power produced and time duration (see Fig. 5.16a). We then quantify the total energy trapped in the arrays, integrating along time the power produced by each piezo disk and summing all the obtained values. It can be seen that, for the trapping case, the energy remains inside the array for a longer period of time (Fig. 5.17), resulting in the highest trapped energy after approximately 28 ms (Fig. 5.17). The total trapped energy at 40 ms is 1.79, 8.74 and 11.03 nJ for the ungraded, metawedge and rotated cells respectively.

Therefore, by utilising rainbow trapping over rainbow reflection, it is possible to harvest more energy along the array, even though the largest local power was achieved by the reflective array; the simplicity of this structure allows the input mode shapes to match the modes of the single rods more effectively than in the symmetry broken arrays. The trapping arrays overcome this apparent downfall through the length of time the energy is trapped along the array, due to the lack of coupling to the reversed waves. Further to this, since the unit cells are more complex (i.e. more rods per cell) there is further scope for including larger amounts of piezoelectric material without compromising the resonances of the rods, e.g. at the base of every rod.

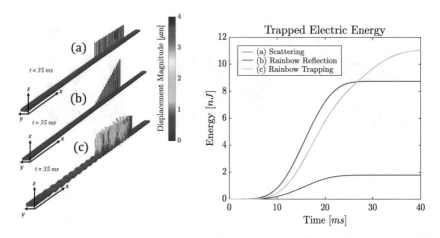

Fig. 5.17 Displacement field along the arrays at time 35 ms for the ungraded (**a**) and the graded (**b**), (**c**) cases, and total electric energy stored in the arrays for increasing time (input stops at 30 ms) [11]

5.4 Topological Rainbow Trapping in Graded SSH Systems

We show now the last multiresonator design proposed in this chapter, based on the activation of proper *topological edge modes*, as in [26]. Topological insulators are systems characterized by protected edge or interfacial surface states between bulk band gaps, due to broken symmetries within a periodic system. Initially studied in quantum mechanical systems [27–29], there has been intensive research translating these effects into classical wave propagation, from electromagnetism and acoustics to mechanics and elasticity [30, 31]; the protected edge modes exploit attractive properties such as resilience to backscattering from defects and impurities, and can show unidirectional propagation. As such the physical phenomena surrounding topological insulators has led to a concerted effort to carry such effects into metamaterial and photonic crystal (and their analogues) designs [32–34]. The nature of the symmetry breaking that creates a protected edge state defines two classes of topological insulators. Active topological materials [35–39] were inspired by quantum mechanical systems exhibiting the quantum Hall effect (QHE), where time reversal symmetry (TRS) is broken through applied external fields [40, 41]. Passive topological systems are based on the quantum spin Hall effect (QSHE), in which symmetry breaking is obtained through spin-orbit interactions (TRS is preserved) [42, 43]. Such systems promoted topological insulator designs in continuum-wave systems based on geometric symmetries breaking to induce topologically nontrivial band gaps [44]. The topological nature of the Bloch bands is associated to peculiar invariants which characterise the geometric phase, i.e. the phase change associated with a continuous, adiabatic deformation of the system; most notably the Berry phase [45, 46], and its one dimensional counterpart, the Zak phase [47]. Concepts based on 2D topological insulators have been translated to wave physics, often defining honeycomb structures [48], which guarantee symmetry-induced Dirac points that can be leveraged to induce edge states at the interface between two topologically distinct media. These have been replicated for waveguiding applications for photonics [49], phononics [50], platonics [51, 52] and acoustics [53]. In other cases, beam splitter designs with square lattices [16, 17, 54] or higher order topological effects, for higher dimensional structures [55, 56] were considered. Despite their relative simplicity, 1D topological insulators are endowed with important features for applications ranging from lasing [57] to mechanical and acoustic transport [58–60]. Motivated by applications in elastic energy harvesting, we highlight a modality of 1D topological insulators, based on the well-established *Su–Schrieffer–Heeger* (SSH) model [61], via the amalgamation of this model with graded structures [13, 24, 25, 62]. Figure 5.18 highlights the main concept, which combines the conventional SSH model with a graded system. In Fig. 5.18a is shown an elastic version of the classical SSH interface, based on resonant rods atop a beam. The interface at which a topological edge mode exists, i.e. where two related geometries meet, is highlighted by the blue dashed line. To the left of this interface are a set of periodic unit cells, of width a, each with two resonant rods set apart a distance Δ_1 from the centre of the cell: we call this structure A. To the right of the interface we consider structures A', that consist of unit cells of the same width, but this time with the rods placed a distance $\Delta_2 = a - \Delta_1$ apart from the cell centre. In an infinite periodic array both these structures are identical as there is merely a translation in the definition of the unit cell. For that infinite array, both structures have the same dispersion curves. However, an edge mode arises at this interface, sometimes referred to as a domain wall.

In Fig. 5.18b a conventional graded metawedge structure [24], like the one previously proposed for energy harvesting [3] is reported. This consists of periodically spaced rods, which increase in height through an adiabatic grading. The utility of such devices is provided by their ability to manipulate and segregate frequency components by slowing down waves which can reach effective local band gaps at different spatial positions [11]. The idea is to combine these two structures, as shown in Fig. 5.18c, to incorporate several, simultaneous, topologically protected edge modes for energy harvesting applications. Such a structure is devised by alternating between primed and un-primed pairs of structures for differing rod heights, as denoted by different letters. We add here piezoelectric materials at the base of the resonating rods, to demonstrate that efficient energy harvesting con be obtained through topological SSH systems; this extends the applications of coupling piezoelectricity with topological insulators [63]. The motivation of coupling with the graded structures, as highlighted in Fig. 5.18, is to enlarge the bandwidth from the single frequency at which the edge mode exist, thereby achieving broadband performance with an attractively compact device.

As previously explained, the graded resonant metawedge [24] allows to locally increase the energy in the resonators, by reducing the effective group velocity of propagating waves. The classical arrangement is that of resonant rods atop an elastic half-space, or elastic beam, as shown in Fig. 5.18b. In this example, the rods adiabatically change in height from one unit cell to the next, generating locally periodic cells; the global behaviour of the device is inferred from the dispersion curves cor-

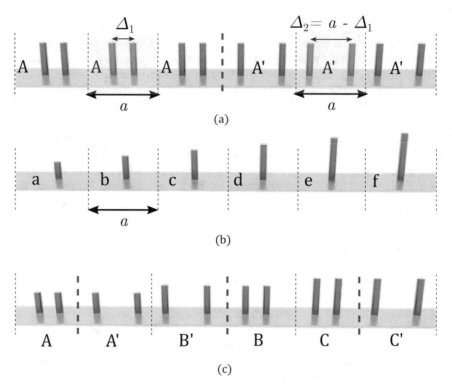

Fig. 5.18 Schematic of SSH metawedge structures. **a** Elastic version of the SSH model, where an interface is encountered between structures A and A′. The unit cells of each are highlighted in black, each shares the same periodicity a, but have rod separation Δ_1 and Δ_2 respectively. **b** Conventional metawedge structure consisting of periodically spaced rods of increasing height, from cell a to f. **c** Amalgamation of these geometries to produce several SSH interfaces for differing rods heights, marked by the dashed blue lines. The green disk at the base of each rod is made of a piezoelectric material which is used for energy harvesting

responding to an infinite array of each rod height [14]. As such, different frequency components encounter local band gaps at different spatial positions.

Similar to rainbow trapping devices [64, 65], the metawedge achieves local field enhancement which can be used for energy harvesting effects [3]. Despite the success, both in design and experimental verification, of a wide variety of effects exhibited by the metawedge and similar structures [13, 25, 62], this simplistic array has reflections, due to Bragg scattering, at the 'trapping' positions. As such, energy is not confined for prolonged periods due to intermodal coupling, and rainbow reflection phenomena is seen instead [11].

The resilience to back-scatter and strong wave confinement that characterise topological systems, makes them attractive candidates for energy extraction; the longer energy is confined to a spatial position, the more energy that can be harvested [11]. This is more efficient for symmetry broken systems, where a lack of coupling to

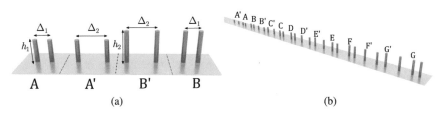

Fig. 5.19 Graded SSH schematics: **a** shows the A-A'-B'-B altering cell structure for heights h_1 and h_2, while **b** shows graded-SSH-metawedge for 7 alternating SSH cells. The green disks at the base of each rod represent a piezoelectric material

reflected waves leads the energy to be more localised; a natural extension of this is to consider topological devices. Indeed, recent designs for topological rainbow effects have been theorised for elasticity in perforated elastic plates of varying thickness, based on topologically protected zero-line-modes (ZLMs) between an interface of 2D square array structures [16], and for acoustic valley phononic crystals [66].

Due to the low dimensionality of the 1D-SSH system, the SSH model provides an optimal arrangement for elastic energy harvesting as there is no propagating component of the edge mode. However, the caveat to this has already been alluded to—this mode only exists for a very narrow range of frequencies. An hybrid graded-SSH-metawedge, based on adiabatic grading of alternating SSH structures, is then designed in order to significantly increases the bandwidth of operation, obtaining true topological rainbow trapping. Due to the strong interaction between the symmetry broken structure and the edge mode, these devices offer an additional benefit of being compact compared to classical metawedges.

Figure 5.19 shows the proposed design, as a zoom of Fig. 5.18, consisting of resonant rods atop an elastic beam. Similar to the motivational mass loaded case, we define structures A and A' to be unit cells consisting of rods of height h_1 arranged in the SSH configuration. The heights of the rods are adiabatically increased every two unit cells, with the arrangement being mirrored: cells with rods of height h_2 follow a B'-B interface. This is repeated along the array. An example of the corresponding A-A'-B'-B geometry is shown in Fig. 5.19a. The advantages of altering the cells in this pairwise fashion (as opposed to A-A'-B-B') come from that, given the grading is suitably adiabatic, there appears to be more cells with the same configuration i.e. on either side of the SSH interface there are two cells with the same structure; conventional SSH interfaces states are typically formed with larger numbers of identical cells either side of the interface. We will show that using this configuration we are able to excite several simultaneously with as little as one of each structure within the graded hybrid device.

Topological systems have been widely proposed as efficient solutions for elastic energy transport, guiding and localization [16, 21]. These concepts offer, amongst others, promising capabilities for energy harvesting, due to the enhancement of local vibrational energy present in the environment. One of the main challenges in elastic energy scavenging, is obtaining simultaneously broadband and compact devices

[1]. Broadband behaviour is usually achieved through nonlinear effects [67, 68] or multimodal response [69, 70], i.e. by exploiting multiple bending modes of continuous beams or arrays of cantilevers. Whilst multimodal harvesting enhances the operational bandwidth, it is usually accompanied by an increase in the volume or weight of the device. This can affect the overall power density of the system as well as the circuit interface, which becomes more complex with respect to single mode harvesters. Conversely it is important to appreciate that multimodal schemes can be well integrated with metamaterial concepts, leading to truly multifunctional designs [3, 71] with enhanced energy harvesting performances.

Here we adopt a multimodal scheme to create a broadband device which is simultaneously compact due to the reduced number of required cells. The device is similar to that in [3], but based on the excitation of local edge modes through the graded-SSH-metawedge geometry (Fig. 5.19). We recall that the physics of these arrays is primarily governed by the longitudinal (axial) resonances of the rods [24] which, along with the periodicity, determine band-gap positions through their resonance. Since the axial resonance frequency of the rod is governed by the rod height [24], by a simple variation of the length of adjacent rods, an effective band-gap, that is both broad and sub-wavelength can be achieved. The addition of alternating SSH configurations introduces frequency dependent positions of localized edge states, that will hence define a true topological rainbow. To quantify the advantages of such designs for energy harvesting, we compare its performance with a conventional rainbow reflection device [3], like the one in the previous sections, composed of equal number of rods, with identical grading angle and quantity of piezoelectric material. The existence of an edge mode is first confirmed by considering two arrays, one only composed of equal rods with constant spacing, i.e. consisting only of structures A (Fig. 5.20a), and another with a transition between regions consisting of structures A and A′, shown in Fig. 5.20b.

Figure 5.20 shows the dispersion relations for a periodic array of cells composed of two aluminium ($\rho = 2710$ kg/m^3, $E = 70$ GPa and $\nu = 0.33$) rods of length 82 mm and circular cross section 3 mm radius atop an aluminium beam. The beam is defined by 10 mm thickness and 30 mm width, and it is assumed infinitely long in the direction of the wave propagation. The unit cell dimension is $a = 30$ mm, with the resonator separation inside the cell as $\Delta_1 = 10$ mm in structure A and $\Delta_2 = a - \Delta_1 = 20$ mm in A′. As expected, the two dispersion relations are identical as can be seen from the paths $-X - \Gamma$ and $\Gamma - X$ in Fig. 5.20b. The colormap of the dispersion curves show the relative polarisation of the rods with green points corresponding to vertical (longitudinal) polarisation, whilst blue refers to horizontal (flexural) polarisation. The Fourier spectra obtained by the scattering simulations exciting at central interface between A and A′, demonstrate the existence of an edge mode. The dispersion curves for both configurations are calculated using Abaqus with a user defined code able to impose Bloch–Floquet boundary conditions. To detect the presence of an edge mode, we excite both systems with a time domain frequency sweep in the range 5–15 kHz, with a source inside the array and located at the interface between A-A′. By inspection of the spatiotemporal Fourier transform of the resultant wavefield, an edge mode clearly appears inside the band gap characterised by an axial resonance.

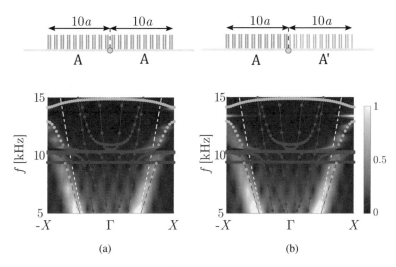

Fig. 5.20 Dispersion relation comparisons for **a** trivial and **b** SSH interfaces. The schematic on the top show the arrangement of rods atop a beam with 10 cells of width a consisting in **a** of structures A-A (purple rods), forming a trivial interface, with **b** showing an SSH interface between 10 cells of structure A and 10 cells of A' (green rods). The Fourier spectrum for this arrangement is computed from a scattering simulation where a source is placed at the interface, marked by the circle and dashed black lines. An edge mode clearly appears inside the band gap defined by the longitudinal resonance of the rods for the SSH arrangement (**b**). Overlaid are the numerical dispersion curves for a perfectly periodic, infinite array of structures A (and simultaneously A') represented by the coloured points, with green points corresponding to vertical polarization of the rod (axial elongation), whilst blue refers to horizontal (flexural motion). The geometrical parameters are such that a = 30 mm with the SSH spacing $\Delta_1 = 10$ mm. The rods have a height of $h = 82$ mm and circular cross section of radius $r = 3$ mm. The beam has thickness $t = 10$ mm and width $w = 30$ mm, and it is assumed to be infinitely long along the array direction

We now address the question of why we decided to alter the height of the resonators instead of the initial spacing Δ_1 in A. In Fig. 5.21 the rationale behind this is shown, which is to ultimately achieve broadband performance. Figure 5.21c, d show the effects of the longitudinal resonance on the position of the Bragg gap in which the edge mode lies; the taller the rod the lower in frequency the axial resonance, which pushes down in frequency the Bragg gaps below it; the geometrically induced band gaps are influenced by the resonance frequencies of the rods, which permits a natural tunability of the devices. Contrary to this, if instead we chose to alter the spacing Δ_1, a much smaller range of edge mode frequencies can be exploited. Given the previous definitions of A, we see that $\Delta_1 < a/2$; for $\Delta_1 > a/2$ structure A is simply interchanged with the primed configuration due to the symmetries of the unit cell, whereas if $\Delta_1 = a/2$ it is not possible to create an interface which distinguishes between the two geometries. As such, there is a narrow range of values which Δ_1 can take. This is highlighted in Fig. 5.21a, b, where the spacing Δ_i is marked with Roman subscripts to avoid confusion between Δ_1 and Δ_2 used in the definition of A and A'. In each case the different Δ_i correspond to altering Δ_1. We alter from $\Delta_b = 14$ mm,

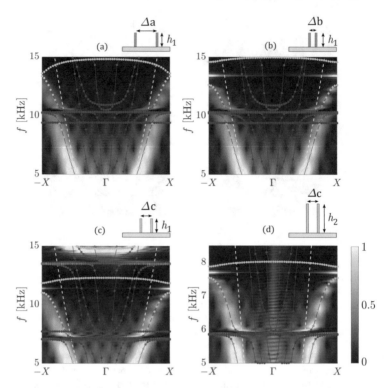

Fig. 5.21 Grading height and spacing: polarised dispersion curves, with longitudinal motion shown by green points, overlaid on the Fourier spectra of the SSH interfaces consisting of different parameters within A and A′. The periodicity $a = 30$ mm and rod radius $r = 3$ mm remain constant throughout. Panels **a**, **b** show the effect of grading the spacing Δ_1 in structures A. There is a small range of values this can take as we are limited by the symmetry of the unit cell. Panel **a** shows the band gap opening for a large $\Delta_1 = \Delta_a = 14$ mm and **b** shows the edge mode for $\Delta_1 = \Delta_b = 7$ mm. Panels **c**, **d** show the effect of grading the height of the resonators; the longitudinal resonances of the rods influence the position of the Bragg gap, allowing for increased bandwidth of the device. **c** has parameters such that $\Delta_1 = \Delta_c = 10$ mm, $h_1 = 100$ mm whilst **d** has the same A spacing with $h_2 = 155$ mm. The edge modes are clearly visible. Note the different scaling on the frequency axis; the gap position has been decreased due to the longitudinal resonance of the rods

which is close to the largest separation where the geometries can be distinguished, in Fig. 5.21a to $\Delta_c = 7$ mm in Fig. 5.21b. These show that the position of the band gap, and hence edge mode frequency, is largely unaffected by the change in spacing. As such for the simplest designs, where there is only one grading parameter, the choice of grading the rod height leads to optimal performance.

We consider a linear graded array of resonators, to be able to provide external excitation and to enlarge the bandwidth, adopting the designs shown in Figs. 5.19 and 5.22a. Thus it is the height of the rods in the pairs of SSH cells which are graded (as opposed to, say, the spacing Δ_1). To quantify the energy that can be stored, we insert PZT-5H piezoelectric disks ($\rho = 7800$ kg/m^3, $E = 61$ GPa and $\nu = 0.31$) of

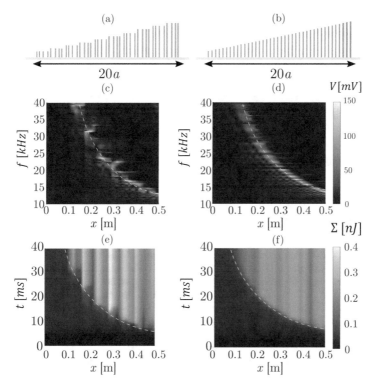

Fig. 5.22 a–b Schematics of graded-SSH-metawedge and conventional metawedge respectively. Open circuit voltage (**c, d**) and total cumulated electric energy (**e, f**) for the graded SSH and conventional metawedge as a function of position along the array

2 mm thickness between the rods and the beam (shown as green disks in Fig. 5.19). Due to the dominant axial elongation in the rod response, we model the piezoelectric coupling by means of the 33 mode piezoelectric coefficient $e_{33} = 19.4$ C/m^2, and constant-stress dielectric constant $\varepsilon_{33}^T/\varepsilon_0 = 3500$, with $\varepsilon_0 = 8.854$ pF/m the free space permittivity. The device is composed of 40 rods with height approximately from 5 to 100 mm and grading angle $\theta \simeq 4.7°$. We compare the SSH rainbow system with a conventional rainbow device, through a steady state dynamic direct analysis performed using Abaqus with open circuit electric conditions.

The infinite length of the beam is modeled using ALID boundaries at the edges [5]. We see rainbow effects in both cases (Fig. 5.22c, d), i.e. spatial signal separation depending on frequency, but the voltage peaks are more localized and with higher amplitude in the SSH case. It is important to notice that this effect is more significant in the steady state regime since a relatively long excitation is required to properly activate the edge modes. Both systems are compared using a time domain simulation with a frequency sweep in the range 10–40 kHz with a source duration of 40 ms. In order to quantify the amount of electric energy stored in both cases, we attach each piezo disk to an electric load of 10 kΩ by means of a user Fortran subroutine integrated with Abaqus implicit time domain integration scheme. The accumulated energy as a function of time is shown in Fig. 5.22e, f. The excitation of the edge modes at discrete frequencies can clearly be seen, with an approximate maximum value of stored energy of 0.44 nJ. For the conventional metawedge, we see that energy is more evenly distributed along space, with a maximum value of approximately 0.26 nJ. This implies that, once the edge modes have been efficiently excited in the SSH configuration, we obtain a local enhancement of approximately 40% of the trapped electric energy when compared to conventional reflective rainbow metawedge configurations.

References

1. A. Erturk, D.J. Inman, *Piezoelectric Energy Harvesting* (Wiley, New York, 2011)
2. R. Ardito, A. Corigliano, G. Gafforelli, C. Valzasina, F. Procopio, R. Zafalon, Advanced model for fast assessment of piezoelectric micro energy harvesters. Front. Mater. (2016)
3. J.M. De Ponti, A. Colombi, R. Ardito, F. Braghin, A. Corigliano, R.V. Craster, Graded elastic metasurface for enhanced energy harvesting. New J. Phys. (2020)
4. N. Elvin, A. Erturk, *Advances in Energy Harvesting Methods* (Springer Nature, New York, 2013)
5. P. Rajagopal, M. Drozdz, E.A. Skelton, M.J.S. Lowe, R.V. Craster, On the use of absorbing layers to simulate the propagation of elastic waves in unbounded isotropic media using commercially available finite element packages. NDT E Int. (2012)
6. J.M. De Ponti, A. Colombi, E. Riva, R. Ardito, F. Braghin, A. Corigliano, R.V. Craster, Experimental investigation of amplification, via a mechanical delay-line, in a rainbow-based metamaterial for energy harvesting. Appl. Phys. Lett. **117**(14) (2020)
7. V.V. Krylov, R.E.T.B. Winward, Experimental investigation of the acoustic black hole effect for flexural waves in tapered plates. J. Sound Vib. (2007)
8. D.J. O'Boy, V.V. Krylov, V. Kralovic, Damping of flexural vibrations in rectangular plates using the acoustic black hole effect. J. Sound Vib. (2010)
9. V.B. Georgiev, J. Cuenca, F. Gautier, L. Simon, V.V. Krylov, Damping of structural vibrations in beams and elliptical plates using the acoustic black hole effect. J. Sound Vib. (2011)
10. E. Riva, M.I.N. Rosa, M. Ruzzene, Edge states and topological pumping in stiffness-modulated elastic plates. Phys. Rev. B (2020)
11. G.J. Chaplain, D. Pajer, J.M. De Ponti, R.V. Craster, Delineating rainbow reflection and trapping with applications for energy harvesting. New J. Phys. (2020)
12. L.G. Bennetts, M.A. Peter, R.V. Craster, Graded resonator arrays for spatial frequency separation and amplification of water waves. J. Fluid Mech. (2018)
13. A. Colombi, V. Ageeva, R.J. Smith, A. Clare, R. Patel, M. Clark, D. Colquitt, P. Roux, S. Guenneau, R.V. Craster, Enhanced sensing and conversion of ultrasonic Rayleigh waves by elastic metasurfaces. Sci. Rep. (2017)
14. V. Romero-García, R. Picó, A. Cebrecos, V.J. Sánchez-Morcillo, K. Staliunas, Enhancement of sound in chirped sonic crystals. Appl. Phys. Lett. (2013)
15. J.D. Joannopoulos, P.R. Villeneuve, S. Fan, Photonic crystals. Solid State Commun. (1997)
16. M.P. Makwana, G.J. Chaplain, Tunable three-way topological energy-splitter. Sci. Rep. **9**, 1–16 (2019)
17. M. Makwana, R. Craster, S. Guenneau, Topological beam-splitting in photonic crystals. Opt. Express (2019)

18. G.J. Chaplain, M.P. Makwana, R.V. Craster, Rayleigh–Bloch, topological edge and interface waves for structured elastic plates. Wave Motion (2019)
19. M. Miniaci, R.K. Pal, B. Morvan, M. Ruzzene, Experimental observation of topologically protected helical edge modes in patterned elastic plates. Phys. Rev. X (2018)
20. M. Miniaci, R.K. Pal, R. Manna, M. Ruzzene, Valley-based splitting of topologically protected helical waves in elastic plates. Phys. Rev. B (2019)
21. M.P. Makwana, R.V. Craster, Designing multidirectional energy splitters and topological valley supernetworks. Phys. Rev. B (2018)
22. D.V. Evans, R. Porter, Penetration of flexural waves through a periodically constrained thin elastic plate in vacuo and floating on water. J. Eng. Math. (2007)
23. Q. Gan, Z. Fu, Y.J. Ding, F.J. Bartoli, Ultrawide-bandwidth slow-light system based on THz plasmonic graded metallic grating structures. Phys. Rev. Lett. (2008)
24. A. Colombi, D. Colquitt, P. Roux, S. Guenneau, R.V. Craster, A seismic metamaterial: the resonant metawedge. Sci. Rep. (2016)
25. G.J. Chaplain, J.M. De Ponti, A. Colombi, R. Fuentes-Dominguez, P. Dryburg, D. Pieris, R.J. Smith, A. Clare, M. Clark, R.V. Craster, Tailored elastic surface to body wave Umklapp conversion. Nat. Commun. (2020)
26. G.J. Chaplain, J.M. De Ponti, G. Aguzzi, A. Colombi, R.V. Craster, Topological rainbow trapping for elastic energy harvesting in graded Su-Schrieffer-Heeger systems. Phys. Rev. Appl. **14**(054035), 15 (2020)
27. J.E. Moore, The birth of topological insulators (2010)
28. M.Z. Hasan, C.L. Kane, Rev. Mod. Phys. **82**, 3045 (2010); Colloquium: topological insulators. Rev. Mod. Phys. (2010)
29. X.L. Qi, S.C. Zhang, Rev. Mod. Phys. **83**, 1057 (2011); Topological insulators and superconductors. Rev. Mod. Phys. (2011)
30. Z. Yang, F. Gao, X. Shi, X. Lin, Z. Gao, Y. Chong, B. Zhang, Topological acoustics. Phys. Rev. Lett. (2015)
31. H. Chen, H. Nassar, G.L. Huang, A study of topological effects in 1D and 2D mechanical lattices. J. Mech. Phys. Solids (2018)
32. A.B. Khanikaev, S.H. Mousavi, W.K. Tse, M. Kargarian, A.H. MacDonald, G. Shvets, Photonic topological insulators. Nat. Mater. (2013)
33. S.H. Mousavi, A.B. Khanikaev, Z. Wang, Topologically protected elastic waves in phononic metamaterials. Nat. Commun. (2015)
34. R. Süsstrunk, S.D. Huber, Observation of phononic helical edge states in a mechanical topological insulator. Science (2015)
35. P. Wang, L. Lu, K. Bertoldi, Topological phononic crystals with one-way elastic edge waves. Phys. Rev. Lett. (2015)
36. L.M. Nash, D. Kleckner, A. Read, V. Vitelli, A.M. Turner, W.T.M. Irvine, Topological mechanics of gyroscopic metamaterials. Proc. Natl. Acad. Sci. USA (2015)
37. A. Souslov, B.C. Van Zuiden, D. Bartolo, V. Vitelli, Topological sound in active-liquid metamaterials. Nat. Phys. (2017)
38. X. Zhang, M. Xiao, Y. Cheng, M.H. Lu, J. Christensen, Topological sound (2018)
39. A. Souslov, K. Dasbiswas, M. Fruchart, S. Vaikuntanathan, V. Vitelli, Topological waves in fluids with odd viscosity. Phys. Rev. Lett. (2019)
40. K.V. Klitzing, G. Dorda, M. Pepper, New method for high-accuracy determination of the fine-structure constant based on quantized Hall resistance. Phys. Rev. Lett. (1980)
41. F.D.M. Haldane, Model for a quantum Hall effect without Landau levels: condensed-matter realization of the "parity anomaly". Phys. Rev. Lett. (1988)
42. C.L. Kane, E.J. Mele, Quantum spin Hall effect in graphene. Phys. Rev. Lett. (2005)
43. B.A. Bernevig, T.L. Hughes, S.C. Zhang, Quantum spin Hall effect and topological phase transition in HgTe quantum wells. Science (2006)
44. M.P. Makwana, R.V. Craster, Geometrically navigating topological plate modes around gentle and sharp bends. Phys. Rev. B (2018)

45. M.V. Berry, Quantal phase factors accompanying adiabatic changes. Proc. R. Soc. Lond. A. Math. Phys. Sci. (1984)
46. D. Xiao, M.C. Chang, Q. Niu, Berry phase effects on electronic properties. Rev. Mod. Phys. (2010)
47. J. Zak, Berrys phase for energy bands in solids. Phys. Rev. Lett. (1989)
48. L.H. Wu, X. Hu, Scheme for achieving a topological photonic crystal by using dielectric material. Phys. Rev. Lett. (2015)
49. A.B. Khanikaev, G. Shvets, Two-dimensional topological photonics (2017)
50. J. Lu, C. Qiu, M. Ke, Z. Liu, Valley vortex states in sonic crystals. Phys. Rev. Lett. (2016)
51. R.K. Pal, M. Ruzzene, Edge waves in plates with resonators: an elastic analogue of the quantum valley Hall effect. New J. Phys. (2017)
52. R. Chaunsali, C.W. Chen, J. Yang, Subwavelength and directional control of flexural waves in zone-folding induced topological plates. Phys. Rev. B (2018)
53. Y.G. Peng, C.Z. Qin, D.G. Zhao, Y.X. Shen, X.Y. Xu, M. Bao, H. Jia, X.F. Zhu, Experimental demonstration of anomalous Floquet topological insulator for sound. Nat. Commun. (2016)
54. M.P. Makwana, N. Laforge, R.V. Craster, G. Dupont, S. Guenneau, V. Laude, M. Kadic, Experimental observations of topologically guided water waves within non-hexagonal structures. Appl. Phys. Lett. (2020)
55. X. Ni, M. Weiner, A. Alù, A.B. Khanikaev, Observation of higher-order topological acoustic states protected by generalized chiral symmetry. Nat. Mater. (2019)
56. M. Li, D. Zhirihin, M. Gorlach, X. Ni, D. Filonov, A. Slobozhanyuk, A. Alù, A.B. Khanikaev, Higher-order topological states in photonic Kagome crystals with long-range interactions. Nat. Photonics (2020)
57. M. Parto, S. Wittek, H. Hodaei, G. Harari, M.A. Bandres, J. Ren, M.C. Rechtsman, M. Segev, D.N. Christodoulides, M. Khajavikhan, Edge-mode lasing in 1D topological active arrays. Phys. Rev. Lett. (2018)
58. B.G.G. Chen, N. Upadhyaya, V. Vitelli, Nonlinear conduction via solitons in a topological mechanical insulator. Proc. Natl. Acad. Sci. USA (2014)
59. Y.G. Peng, Z.G. Geng, X.F. Zhu, Topologically protected bound states in one-dimensional Floquet acoustic waveguide systems. J. Appl. Phys. (2018)
60. Y.X. Shen, L.S. Zeng, Z.G. Geng, D.G. Zhao, Y.G. Peng, X.F. Zhu, Acoustic adiabatic propagation based on topological pumping in a coupled multicavity chain lattice. Phys. Rev. Appl. (2020)
61. W.P. Su, J.R. Schrieffer, A.J. Heeger, Solitons in polyacetylene. Phys. Rev. Lett. (1979)
62. E.A. Skelton, R.V. Craster, A. Colombi, D.J. Colquitt, The multi-physics metawedge: graded arrays on fluid-loaded elastic plates and the mechanical analogues of rainbow trapping and mode conversion. New J. Phys. (2018)
63. S. McHugh, Topological insulator realized with piezoelectric resonators. Phys. Rev. Appl. (2016)
64. K.L. Tsakmakidis, A.D. Boardman, O. Hess, 'Trapped rainbow' storage of light in metamaterials. Nature (2007)
65. J. Zhu, Y. Chen, X. Zhu, F.J. Garcia-Vidal, W. Yin, W. Zhang, X. Zhang, Acoustic rainbow trapping. Sci. Rep. (2013)
66. Z. Tian, C. Shen, J. Li, E. Reit, H. Bachman, J.E.S. Socolar, S.A. Cummer, T.J. Huang, Dispersion tuning and route reconfiguration of acoustic waves in valley topological phononic crystals. Nat. Commun. (2020)
67. F. Cottone, H. Vocca, L. Gammaitoni, Nonlinear energy harvesting. Phys. Rev. Lett. (2009)
68. A. Erturk, D.J. Inman, Broadband piezoelectric power generation on high-energy orbits of the bistable Duffing oscillator with electromechanical coupling. J. Sound Vib. (2011)
69. S.M. Shahruz, Design of mechanical band-pass filters with large frequency bands for energy scavenging. Mechatronics (2006)
70. M. Ferrari, V. Ferrari, M. Guizzetti, D. Marioli, A. Taroni, Piezoelectric multifrequency energy converter for power harvesting in autonomous microsystems. Sens. Actuators A: Phys. (2008)
71. C. Sugino, A. Erturk, Analysis of multifunctional piezoelectric metastructures for low-frequency bandgap formation and energy harvesting. J. Phys. D: Appl. Phys. (2018)

Conclusions

In this book, a general overview of elastic wave manipulation mechanisms in the perspective of energy harvesting is provided. After having introduced the world of metamaterials, distinguished from inhomogeneous media, a comprehensive review of energy harvesting technologies based on structuring materials and metamaterials is proposed. Basic concepts on wave propagation in homogeneous and inhomogeneous media are then introduced for both periodic and aperiodic structures through one-dimensional lumped models. Starting from the classical theory of wave propagation in homogeneous beams and periodic spring-mass chains, the concept of band gap is interpreted through energy considerations and Argand diagrams. Periodic and aperiodic structures with equal global mass and stiffness are compared. It is demonstrated that similar attenuation levels can be obtained perturbing the distribution of masses and springs. Depending on the level of perturbation, it is possible to anticipate the attenuation trend but this is usually accompanied by transmission peaks inside the attenuation region of the corresponding periodic structure. A physical interpretation of local resonance is provided through analytical lumped models. It is demonstrated that the efficiency of the local resonance band gap is not only affected by the resonating mass, but strongly depends on the stiffness of the main structure with respect to the one of the resonator. Transmission analyses show that increasing the stiffness of the main structure with respect to the one of the resonator, a sharp resonance and antiresonance is obtained without attenuation, as also demonstrated by the imaginary part of the dispersion curves. The problem of wave propagation in elastic continua, with specific reference to plates and half-spaces with resonators is then investigated. It is demonstrated that strong interaction is obtained between the A_0 Lamb mode and the longitudinal resonance of the rod. The concept of grading is introduced as a way to obtain a band gap that is both broad and subwavelength. Specifically, it is noticed that exciting the plate with an A_0 Lamb mode from short to high resonators, the waves are slowed down until the band gap opening (with wavelength reduction), while in the opposite direction they are immediately stopped and reflected backward. In the first case, strong amplification can be obtained in the resonator in the position in which

© The Author(s), under exclusive license to Springer Nature Switzerland AG 2021   121
J. M. De Ponti, *Graded Elastic Metamaterials for Energy Harvesting*,
PoliMI SpringerBriefs, https://doi.org/10.1007/978-3-030-69060-1

the wave stops. Similarly, the problem of wave propagation in elastic half-spaces is considered, with the definition of the so called *metawedge*, i.e. an array of resonators able to provide rainbow effect or mode conversion depending on the direction of the incident Rayleigh wave with respect to the array grading. Reversed mode conversion from surface Rayleigh to pressure and shear bulk waves is demonstrated leveraging on the Umklapp phenomenon. Piezo augmented arrays of resonators able to dramatically increase the energy available for harvesting, and the operational bandwidth are then presented. After an introduction to piezoelectricity, a *rainbow reflection* device made of an array of rods with increasing height on an elastic beam is considered. It is demonstrated that the metasurface provides a strong amplification of the electric power, and when the harvester is embedded in the metasurface, due to the lower group velocity of the waves interacting with the array grading, the power generation peak occurs at a later time. Broadband performances can be obtained placing more harvester in the array. A second design, based on *rainbow trapping* is proposed in order to avoid reflection of energy, thus increasing the harvestable energy in time. This is done designing a symmetry broken array where accidental degeneracies are lifted, resulting in zero group velocity modes inside the first BZ. For a 40 ms input duration, the rainbow reflection array stores 84% more energy than the ungraded array and 20% than the metawedge. Finally, a multiresonator device based on the activation of topological edge modes in a *graded Su–Schrieffer–Heeger (SSH) system* is shown. The energy harvesting enhancement provided by the SSH metawedge is quantified comparing its performance with a rainbow reflection device through numerical (FEM) simulations. This shows an increase of 40% of the total transduced energy.

In conclusion, we have demonstrated potential advantages in using rainbow-based metamaterials for energy harvesting. Rainbow reflection devices allow to slow-down waves, increasing the interaction time between the wave and the harvesters. Strong energy localization is obtained, providing high local values of the electric power in the position associated to the band gap opening. However, due to the zero group velocity mode achieved at the edge of the first BZ, high reflections exist. Enhanced transduced energy along time can be obtained exploiting rainbow trapping in symmetry broken arrays where accidental degeneracies are lifted resulting in zero group velocity modes inside the first BZ. In this way, due to the lack of coupling with the reversed waves, energy remains more time in the vicinity of the harvester, feeding more energy to it. Finally, the resilience to back-scatter, strong wave confinement, and robustness to defects that characterise topological systems, makes them attractive candidates for energy harvesting. The definition of SSH rainbow systems allows to increase locally the peak electric power, as well as the total transduced energy in time.

However, many issues are still to be faced and several future developments of this work are possible. Even if the efficiency of the proposed designs has been verified numerically and experimentally, a problem of manufacturability, especially at microscale, still exists. On the other hand, the presence of several resonators is accompanied by an increased volume or weight of the device, reducing its overall power density (power/volume or power/weight). Parallel to this, an important issue to be faced is to avoid mode shape dependent voltage cancelation and the cancelation due to the phase difference between different harvesters. The integration of rectifiers in the circuitry would allow for the full exploitation of these multiresonant designs, avoiding the charge cancellation. Other issues concern the possibility to use these devices, especially at microscale, for low frequency input spectra. Solutions based on simpler planar geometries, as well as frequency up conversion mechanisms have to be seriously considered to overcome the problem of manufacturability and frequency mismatch between the metasurface and the input source. We finally emphasise that the proposed results should be considered as preliminary proof of concepts able to show the advantages of using graded metamaterials for energy harvesting. Microscale applications, based on the theoretical concepts developed here, should consider other constraints given by the required size of the device, manufacturability issues, and realistic input spectra.

Printed in the United States
By Bookmasters